Born in 1947, on a farm in the White Highlands of Kenya, Roger attended school in Bournemouth, UK, and St David's College in South Africa.

He started on the mines near Johannesburg when he was nineteen years old. He has worked on gold, platinum and copper mines all over the world.

Roger married in 1968 and is the father of four children by his first wife, Sharon. She died of cancer in 1993. He has since remarried and lives with his wife, Cynthia and their young son Gordon, on a forty-foot boat near St Ives in Cambridgeshire.

This book is dedicated to the many workers that went underground and gave their health and life to the pursuit of wealth, a wealth that was for them a monthly pittance, but for the government of the time supported an apartheid ideology.

Roger Russell

VISIBLE GOLD

AUSTIN MACAULEY PUBLISHERS
LONDON * CAMBRIDGE * NEW YORK * SHARJAH

Copyright © Roger Russell 2024

The right of Roger Russell to be identified as author of this work has been asserted by the author in accordance with sections 77 and 78 of the Copyright, Designs and Patents Act 1988.

All rights reserved. No part of this publication may be reproduced, stored in a retrieval system, or transmitted in any form or by any means, electronic, mechanical, photocopying, recording, or otherwise, without the prior permission of the publishers.

Any person who commits any unauthorised act in relation to this publication may be liable to criminal prosecution and civil claims for damages.

All of the events in this memoir are true to the best of author's memory. The views expressed in this memoir are solely those of the author.

A CIP catalogue record for this title is available from the British Library.

ISBN 9781035844333 (Paperback)
ISBN 9781035844340 (ePub e-book)

www.austinmacauley.com

First Published 2024
Austin Macauley Publishers Ltd®
1 Canada Square
Canary Wharf
London
E14 5AA

Table of Contents

Author's Note	9
Introduction	11
Part One	**19**
Shift…	*21*
New Friends	*28*
The First Time	*40*
Malawi	*54*
Mud Rush	*57*
The Old Man in the Rock	*63*
Part Two	**95**
The Creation	*97*
A Pocket Full of Stardust	*101*
In Pursuit of What?	*109*
Part Three	**119**
Majas	*121*
We Are Here! (Tina Kona)	*132*
Strike	*144*
A Touch of Greatness	*194*
Postscript	197

Author's Note

Klerksdorp is a small town in South Africa where gold has been mined for so long that no one knows for sure when it all began. It began for me in 1968. It began when I enrolled in the South African School of Mines and became part of a unique culture. As with most beginnings, it was also an end, the end of childhood and the protection of innocence.

The futility or effectiveness of our lives is often measured by an evaluation of the change we have created in the world around us. Change within ourselves has little value unless it makes an impression on a world that exists into a future beyond our own. This future could be mankind's continued existence or some sort of spiritual eternity. Whatever the future turns out to be, we, by the nature of things, can never be assured of a continued role in it.

The striving that we do, we do in faith, hope or charity; our human frailty using any one of the three as haphazardly as chance dictates. We believe that the results of our actions will assure us of a place in human memory or in an eternity of some kind. Perhaps the one is necessary for the other.

There is, of course, a large part of the human race that seems to have no intention of creating anything permanent at all. For this unfortunate group, there is no future, and they enthusiastically build castles in the sand, enshrining themselves in turrets of greed and self-justification. They laugh today and sing, gazing down on the rest of us from tall, seemingly substantial turrets held together by nothing more than water. Tomorrow, the tide has come and gone, and all is as it was before. The tragedy is that, sooner or later, it begins to look as if they did in fact have the answer.

I write a great deal and have always believed that writing should either pose questions or provide answers. I have found few answers but many questions in what follows. Many of the people that I hope will come to life between these pages acted as if they knew the answers. Their stories and mine are authentic. I have, however, used my imagination to name them and the places where they worked. Their stories do not reflect intensive research; neither do they represent the feelings or memories of others.

The stories are my memories and as such are tempered by my philosophies and coloured by my vision. The intention was never to present the reader with an investigative account but to rather illuminate and present the courage, desperation, pettiness, and camaraderie of an industry. An industry that seems to me deserves to be understood rather than protected or destroyed.

You do not survive on the mines if you do not have one kind of strength or another. I do not suppose that I can say that I really succeeded on the mines in any special way and the person that I was at the beginning did not survive; I came away changed, both by the many people that I met and the society which they have created. These people, both brutal and missionary, have all a strength and it is to them that I dedicate this book; a salute to the roles that they play in a very real and dangerous world.

Introduction

The white man's history claims that the earliest report of gold in that part of the African continent, destined to become the richest reserve in the world, was that of a certain Jan Blank. Jan claimed to know of a mountain of gold in the South African interior in 1652. Jan van Riebeeck considered these claims childish and was possibly correct, but like many childish claims, they had some substance. This reluctance on the part of the authorities to believe reports of findings was to earmark the early stages of the South African gold industry.

Gold, however, has the ability to attract the attention of people who believe in chance and, thanks to the vast numbers of such hopefuls, eventually the payable presence of gold was firmly established.

That gold is one of the great motivators of humanity goes without saying; the need of financiers to own gold has driven exploration, fuelled conquests, and justified the abuse of primitive peoples. Recent history of such enterprise has seen famous goldfields opened and exploited across the globe. California, Alaska, Australia, South America, and Russia are names that conjure up pictures of frenzied activity, death and wild pleasures. Men have streamed across the continents to get rich, taking their passion and greed with them.

The South African gold mines have one singular quality that has differentiated them from every other strike in history—they have never run out. Whereas, the entire Witwatersrand field employed approximately forty thousand blacks in 1895, this number constitutes the strength of a single mine now operating in the Klerksdorp area. Gold is still mined in many other fields all over the world. But if they can be said to be middle aged or dying, then the South African fields are still shaking the bars of the cot and screaming lustily to be placed upright on the floor.

Once upon a time, there were indeed mountains of gold, mountains that were slowly eroded and deposited on the floor of a huge basin. This basin now hosts what is known as the Witwatersrand Super group. The basin, 320 kilometres long and 160 kilometres wide, lies under a large portion of the Transvaal. In section, the basin outcrops on its western rim in the vicinity of Klerksdorp and in the north along the Witwatersrand (Ridge of white waters). At the moment, we mine along the rim. As we get deeper and closer to the bottom of the basin, it is possible that the deposits will get richer.

Without a doubt, there is still a great deal of gold; man must only supply the ingenuity and effort, the reward is as substantial as the earth itself.

To describe a gold mine is difficult.

It is an immense technical complexity and an even greater social phenomenon. On approaching a modern shaft, you are struck by the amount of activity and facilities with which this relatively tiny hole in the ground has surrounded itself. Aside from the concrete tower which houses the headgear and its immediate family of structures, there are numerous other buildings and services which require introduction and explanation.

A refrigeration plant which supplies chilled water in an attempt to cool the temperatures of the air underground from as high as 50°C down to operating temperatures of about 34°C. This particular sophistication allows the mines to go deeper. If current experimentation in the use of crushed ice is successful, then they will be able to go deeper still.

A hostel, housing sometimes up to three or four thousand permanent residents.

A change house, administration buildings, personnel offices, and a huge store complex.

A dressing station for the immediate attention to accident victims as well as daily medical services for the sick, lame and lazy. These and other facilities are only some of the satellite organs that the courtesan nurtures within herself in her lust for life.

A shaft is the magic passage through which the mine above nurtures its gloomy and delinquent foetus below. An umbilical cord of steel and concrete, it feeds and is fed, for the arrangement only lasts as long as what comes up has more value than what goes down. An unfortunate reflection on man's value to man; because much life goes down that is consumed, but what emerges has no life at all.

On such a shaft, three thousand or more black employees will go down the mine on a daily basis. Possibly hundred and fifty white supervisors will accompany them. These men need tools and equipment, materials and services.

They need air to breathe and light to see by. All this must be transported down a shaft 3,000 to 4,000 metres deep and 12 metres wide. Everything must come up the same shaft. The cage, a triple-decker lift for transporting men, will run continuously for nearly two hours to place the teams of miners on the levels where they will spend their shifts. The same will happen in the early afternoon to remove them.

Going down the mine is less stressful than coming out of it. The men are in no hurry and shuffle along the walkways leading to the bank with little urgency. Coming out of the mine is a little different; queuing to get the early cages is a sensitive issue and has often erupted in violence.

A black labourer who has spent his shift in extreme conditions, exhausted, wet and dirty, must stand back after queuing for nearly an hour and watch a near empty cage transport ten or fifteen whites to surface. These whites arrive at the shaft cool and clean as if they had not been underground at all. Without regard to the long lines of blacks, they demand and get the first cage to arrive. When rational, supervisors as well as the majority of the blacks recognise that this tradition will not easily be set aside.

But at the end of a hard shift, in the reality of the immediate, this is not always easy to see. The shafts have seen wildcat strikes with blows struck and men injured because some tired and frustrated machine boy has decided he will not keep his mouth shut or his hands to himself.

A black man's day is physical and strenuous. He is strong but fatalistic; his circumstances demand that he subjugates himself to the whims of a minority who hold court through the might of technical superiority. A child of the grasslands and skies of the veldt, he hands over his fear and scepticism with simplistic faith to the ability of the industry to ensure his survival. Later, when he is clever and experienced, he will become authoritative and will say no when he thinks no and yes when he thinks yes. His perceptions and sense of justice will have been developed and he will sometimes voice them.

The white man's world is very dependent on his own attitude to life. Down the mine, he is away from immediate authority and stamps his own personality on his workers and his job. Many miners are hardworking and well-respected by their work crews. But for the most part, miners are a tough and callous group who work in the darkness because they have to. Circumstance placed them there and they will not or cannot find the courage to leave, it is all they know.

A miner's job is to ensure that the conditions are safe to work in and to personally deal with the application and control of explosives. His shift boss has a more legal and administrative function. He is supposedly the watchdog of the mine and is ultimately responsible for the recording and maintenance of controls. His time underground is less than that of his subordinates and he spends the afternoons on surface planning and organising the support his miners need.

The shift boss answers to management for accident rates, waste, lost production and high costs. Together with his mine captain, he is probably the man in the industry with the least training and most responsibility.

He is underpowered and over-blamed.

He works in an office with the other shift bosses in his section. This office is normally situated en route from shaft to gate. The windows face the walkway and they become the place where audience is granted to workers and miners alike. Between two and four o'clock in the afternoon, this interface becomes a turmoil of men, black and white, still stinking from hard labour in confined and hot conditions, who argue, push and shout in order to get their particular problem resolved and the day finally over.

One shaft might support four or five mine captains with five or six shift bosses reporting to each of them. A shift boss will supervise two miners and each miner, four to five team leaders and up to sixty labourers. A mine captain's section will number anything from two hundred to six hundred people producing up to 30,000 tons of ore each month. The mine captain coordinates planning and relays management feelings and demands to his staff. Contact between him and the underground workforce is one step removed. He is not truly part of the workforce.

He has placed his hands firmly on the ladder leading to comfortable offices and divorced pontification. Around these key positions, the rest of the mine flutters its wings and serves.

Personnel departments hire and fire, train and explain, they process and analyse grievances upwards and control disciplinary action downwards. Survey, ventilation and geology departments are at the beck and call of the mining department; they install markers, analyse conditions, mark out directions, assess grades and measure production. The engineering department is nearly as powerful as the mining team and keeps the proverbial wheels turning, but also, in the long term, dance to the tune that production pipes.

All these departments, their men and their machinery are positioned and operate around a small circular hole in the ground. They fight, squabble, cry and laugh, work, and get drunk with equal ease and regularity within a few metres of a dark and almost bottomless pencil-thin, concrete tube in the earth.

Some shafts are old and square; some are modern, round or oval. Some are twin systems, some go straight down and some go down at an angle.

They all serve one purpose and have this in common; they are great roadways connecting deep hidden worlds of strange and alien traditions to civilised facades on surface. There is no greater distance than that from the surface of the earth to just below it. Science fiction cannot devise a more staggering time warp or culture leap as that from collar elevation to the first working level of a mine. The environment down there is artificial, great fans boost the air along tunnels and drives.

A complicated system of doors and regulators force it roaring down one tunnel and filters it hot and heavy down another. You can walk along, comfortable, cool and easy until you pass an intersection, on the other side of which you hit an almost solid mass of thick, hot, damp air that lifts your pores, pricking with the slow start of sweat. It weighs down your energy, wiping away good intentions like raindrops from a windscreen. What comforts there are, you carry with you; light that exists everywhere, even on the darkest night; rides piggyback on your head in a mine.

If you don't install it, or carry it, it simply does not exist. Stand still and switch off your cap lamp. The darkness is total, it is complete. It lives and breathes all around you; no matter how long you wait, you will never get used to it. You will never be able to see anything because there is nothing to see. The miraculous cluster of nerve endings in the back of your eye lies waiting, deprived of all sensation.

The lord black rules forever down there and you swim in his world in a small bubble of your own making. Without light, without a world; without air, without life; without food and water, without hope and without other men, so alone and so far away. Man, on surface is a human being, but underground, an alien. He does not belong, is not wanted, and not even really tolerated. A miner thinks of himself as dominant and aggressive but the ground he works in is sly and subtle.

The great wonder of his painstaking and dedicated pillage of the earth lies not in that the earth is torn aside or destroyed but in that the galleries and rooms have been neatly cut and bored only where necessary. Only that which is required to be stripped is stripped. All that is taken is what is needed.

It is a wonder that the earth even knows what is going on, but she does, and she protests. Aside from the occasional stupidity of the odd miner who brings down loose rocks on his own and other people's heads, the great mother herself shakes and grows impatient. Sometimes, with little warning, she reaches out and slaps the tunnels around. She takes hold and with a flick of the wrist, collapses excavations, shakes massive rocks loose for kilometres in all directions. She buckles rail tracks and pipes with ease, destroying years of work in moments. When she does this, men die; they are broken and buried with contempt.

Afterwards, those that are left scurry around like ants and rebuild what they can. They cannot stop her, so simply carry on. They will never withdraw; as long as someone is willing to pay, there will be those who climb into a steel box and drop, hurtling downwards into that other world. That strange and troubled place of fists and steel, of death and deep brotherhood.

The difference does not end with bad air and falling rocks; it is propagated in relationships and social structure. It is a false and unnatural place. Authority figures are imposed by the other world, the surface world. An ability to shine in the office or at a party is just as important as an ability to perform underground. A man will find himself reporting to another whom he does not know or even respect. Miners, shift bosses and mine captains are sometimes inadequate, sometimes over aggressive but they are there and must be obeyed.

Sections produce equally well; one under the control of an animal the other under the direction of a philosopher. The system is what counts and the person who can undermine that system is rare and soon disposed of.

Brutality is part of it; beatings, assaults, mockery and insults occur on a daily basis. Some cases come to light and are handled in that upper world, but most are private, things belonging to the place, the people and the circumstances. No one reports them, not even the victims. Position is important, privileges of rank are visible everywhere; bicycles that run on the rails for the miner or shift boss, a pikanin (young boy) to carry iced water and run messages.

Sometimes there is an elaborately constructed chair at the beginning of the working place or at the underground store for the sole use of the miner. It is a throne; men go underground, some to serve, others to be served.

All of them survive by complying with or utilising the unwritten laws as they deem fit.

They are a strange race, these miners, but to be one of them is special; to have secured a place amongst them means something. To be one of them is to know, if I am in trouble they will come from everywhere and dig. No matter how dangerous it is, there will be someone. No matter what the colour of my skin is, no matter what my place is in the ranks; sweat, blood and lives will be spent for me. To be one of them is to know that broken regulations will be hidden, dangerous shortcuts covered up, even stupidity or pettiness kept hidden down below in the world in which it belongs.

To be accepted is to be made a brother, a separate person with relationships and family ties on surface but with stronger and closer ties in the depths of the earth.

A mine often derives its personality from the circumstances of the reef. In some mines, conditions are good. The air is cool and the rock solid. They exist, but most mines are situated in areas of stress and the depth at which mining takes place lends force to an already unreliable and dangerous formation. Heat fuels irritations and the gold is dragged from the rock by sheer obstinacy. These places have no whitewashed waiting places or neat orderly tunnels, they are hellholes of frustration and violence.

They are characterised by highs: high production figures, high morale and high accident rates. Strong men with strong stomachs support tough shafts. Ordinary men lead better lives on better mines but somehow fall short of a standard by which miners rate themselves. Often when they sit and talk, they will tell one another of a time when they worked on such a shaft and how it was. They describe dangerous worlds, worlds of fierce commitment and sudden tragedy. Worlds shaped by circumstance, their parasites and their victims, to infernal designs drawn God knows where, by God knows whom.

Part One

Shift...

When you step out of the cage and onto the level where you work, the shaft and its connection to the upper world get left behind. It becomes only a promise; something that awaits you at the end of the shift. Ahead of you lies the day, an unopened Pandora's box of drama and discomfort.

Some working places are so far from the shaft that you must board a train to get there in any kind of sensible time. If not, then you walk. It is not unusual to have to walk 3 or 4 kilometres to get to your crews. Dressed in thick overalls, with your cap lamp and its battery, perhaps a bag containing lunch and cold or frozen water, it can take anything up to an hour to arrive at the place that is your particular responsibility. You set off, striding along the rail track that runs down the centre of the tunnel.

Heavy gumboots don't make it any easier and neither does the careless spacing of the sleepers. Riding the ore trains is forbidden and by the time that you have waited for one to stop and start its way through ventilation doors and sidings, not really worth it anyway. As you leave the shaft behind, the tunnel becomes lonely and bare: you walk and walk between three supposedly square walls of rock. In order to keep the rocks in these walls stable, the face is studded with anchor bolts and sometimes laced with wire mesh and cables.

You pass black tunnel mouths leading to other places that stay black holes because they do not concern you and have their own tiny communities that serve them. Eventually, as you near the section that belongs to you, the world starts to expand a little. You start to meet people and find evidence that work is in progress.

An old man stoops over a mud-filled drain with a shovel. As you approach him, he straightens and you greet him, "Hello, mdalla (old one), how are you?"

"Hello, lnkosi (chief). Mina kona (I am here)."

He is too old and sickly to perform his normal job and is content to wait for retirement. You have allocated a section of the tunnel to him to look after, and it has become his personal property. He spends his day keeping it clean, painting where he thinks it necessary and yelling at the younger workers when they intrude or dirty the area with their carelessness. A little praise and direction are all that is needed to keep him carefully concerned for the rest of the day.

Further in, a well-lit area indicates a centre of activity and you reach the section store. This is a focal point for planning and organising the job. Key workers know when you will arrive and when you take your breaks, and they expect to find you there at those times. Here, the drillers collect and exchange their drill steel. Miner's assistants fetch paint and exchange worn equipment. If there is a telephone in your section, it will be here, as will be the first aid station. If you have arranged to meet anyone, it will be here; if not, you will question the store boy to establish who and what has passed by this place.

You will most likely stop and rest. It is a place where you will take out your notebook and plan the shift. You might have a last drink of cold water before hanging your bag and coat in the store as you prepare to go into the working place and face the day. It will probably be the last piece of sanity you encounter for some hours. You leave it with reluctance and some trepidation of what is to come. It is not that all shifts underground are bad, it is just that they so easily can be.

Most gold reef ranges from a centimetre to several metres thick and sits between the layers of rock like the cold meat in a sandwich. It is nearly always mined from its underside so that as it is broken up and removed from the face it can be dropped into the tunnels below it. The development of tunnels to reach the reef is known as access development and takes place on several different levels.

As the shaft is sunk, stations are developed at predetermined depths and these serve as the starting points for the tunnels that are driven through the rock to reach the reef horizon. Haulages normally run along behind and below the reef and never actually intersect it. From the haulages, crosscut tunnels are turned off and mined straight at the reef until they intersect it. From this intersection, a raise is mined at an angle along the reef and continues up and down the reef until it intersects the level above and below it. Working panels will be cut along the length of the raise and each panel will be fitted with its necessary machinery and services in a process known as ledging and equipping.

We finally have a working tunnel from one level to the next. This becomes the centre of the mining effort. As you climb up its steep floor, you are climbing up the slope of the reef, which is mined away from you to the right or the left. You may enter the raise from the level above your working panels or from the level below. It is always easier to climb down, so you may prefer to enter on the level above but as most of the action involving your workers occurs below the working places, you will probably have to go in on the lower level.

From your store in the haulage to the bottom of the raise connection, it is not far, perhaps only 600 or 700 metres. But it will take you as long to travel as the 4 kilometres you walked in from the shaft in the morning. The tracks in your crosscut are in poor condition; the constant stream of water flowing from the mining operations undermines the ballast and the overloaded ore trains press the sleepers in and out of the never-ending mud and slush. Pumps and drains continually block or fail and the damp soggy conditions add to the humidity, making the air not only hot but destroying its ability to cool the body.

All along the length of the crosscut are box fronts. A box front is the steel structure at the bottom end of a rockpass, which goes up at an angle of approximately 60º all the way to the working face. At the top, it has an iron grid to stop rocks above a certain size from falling into it, and at the bottom, the box front has a steel door that stops or regulates the flow of ore. The people in the working places above tip the mined ore into the ore pass by dragging it there with a scraper. The box acts as a storage place until an ore train arrives at the boxfront below to bleed off the rock.

A bossboy, or one of the loco crew, climbs up onto the platform and siphons off ore into the hoppers positioned below. Where the tracks pass under the box is always a mess. Spillage caused by over filling the hoppers climbs up the walls on either side of the tunnel. The drains are always blocked here. Dams of dirty water stretch metres into the tunnel ahead of each box area. Sometimes when the hoppers are being filled, you cannot pass and must either climb over the length of the train or stand and wait until the long process of filling all the cars is complete.

The rocks inside the ore pass always seem to be bigger than the steel door. One of the most unpleasant jobs underground can be the slow and arduous coaxing of awkward shaped chunks of granite through an aperture too small to pass them. Steel levers and hammers, a great deal of sweating and swearing, and often the use of explosives is required to keep the rock moving.

The crosscut is not only in use for the outward movement of rock, it is also the only way that your mining supplies such as support timber, explosives and drilling equipment can reach the workface above. The daily influx of material cars must fight its way upstream against the flow of ore on its way out. The timber arrives on flat cars and is unloaded anywhere that there is space in the crosscut. In an already limited excavation, timber overflows much as the spillage does into the drains and onto the tracks. It easily becomes waterlogged and heavy and, as is the way of the world, an unpopular item to those who must carry it.

The timber at the bottom of the pile remains in the mud and adds to the clogged and awkward conditions. Your mood, after traversing this confusion of industry and negligence, becomes tense and irritable. The situation is far from what you would like it to be and you have not, in the couple of months that you have been there, been able to do much to change it. Your budget is almost non-existent and what there is of it seems to go on immediate production. There is no money or labour to fix or replace equipment that has been salvaged and reclaimed so many times that it is worse than useless.

Your production figures are being maintained but only just. Any small delay could put you behind with little hope of making up what has been lost. You climb over the obstacles, wade your way through the mud. You shout angrily at the bossboy who was supposed to pack timber further down the crosscut where there was at least some available space. The bastard has unpacked all of it where the material cars were stopped by some boys working on a burst water pipe…so more water, more soggy timber. He is sullen and answers insolently that he has other work to do and cannot waste time waiting for the tracks to clear.

When you reach the broken pipe, it has been repaired with a piece of rubber and a clamp but is still spraying a steady stream of water into the sidewall. The crew that repaired it is packing up to leave and you question the leader.

"Paulus, what happened here?"

"The pipe was broken, baas, I fixed it."

"Yes, I can see that but why did it break? When did it break?"

He shrugs. "I don't know, it was broken on nightshift?"

"So nobody has done any drilling yet?"

"No, baas, no drilling anywhere in the section."

You look at your watch: it is just after ten.

"Has the Mlung (white man/miner) marked off the holes yet?"

He shrugs again. "I do not know."

You know. You know that the work is seriously behind and that you will lose production if something is not done quickly. You will have to take all the spare labour you have just allocated to cleaning up the crosscut to prepare and drill the faces in time. On the other hand, you could perhaps leave two panels and concentrate on the others, using the spare labour tomorrow to make up lost production. You could do this; you could do that. You need to talk to the miner.

At the reef intersection, there is a monorope installation. It is an electrically driven drum that moves a wire cable. This wire cable travels from the drum, around pulleys and wheels and up the raise to the next level, then back down to the drum in a continuous loop. At the drum, the timber is tied to the cable with a slip knot which, when it tightens, snatches the timber away and carries it up the raise. Where the timber is needed, another worker waits to cut it loose. Today, at the drum, the situation is chaotic; because everything is late, there are at least four blacks using the monorope at the same time.

You push through, climbing awkwardly over piles of bricks, loose timber and explosives. There are some broken pockets of cement and a burnt-out winch motor that has been there for a week despite repeated requests to the electricians to do something about it. Unfortunately for your senior bossboy, he appears, coming down the travelling way as you are about to go in. He stoically weathers your anger and frustration and agrees to fetch labour from the crosscut to clear the mess around you. You tell him to send someone else and instruct him to follow you back into the working places. Today, anger and not objectives will manage your job.

A raise connection is normally mined about 3 metres high and 2 metres wide. The mining is planned so that the reef is carried along the top of the tunnel just below the roof, which is called the hanging wall. Reef varies from place to place but is commonly a band of almost black rock full of little white pebbles. You cannot see the gold, it forms part of the dark band. In very few places where the value is high, the rich yellow can sometimes be seen, but it is rare. To start mining from the raise connection outwards, a slot is blasted out of one side of the raise just under the hanging wall.

This slot is about a metre high and is mined until it is approximately 2 metres deep. Into this space, timber is packed to give support to the roof and a similar exercise is done on the opposite side of the raise. Now to walk up the raise is to walk up a gully about 1.5 metres deep with a timber filled ledge cut into the rock on both sides of you. These ledges become the working faces and the faces are moved into the rock away from the raise. Every 20 to 30 metres along the raise, a gully is mined, which keeps just ahead of the face.

The rock that is blasted on the face is scraped down the length of the face to this gully and then back along the gully to the raise. At the intersection of the gully and the raise, there is an ore pass into which the rock is tipped. The dayshift does the blasting and the nightshift the scraping. If they are good, then the gullies are clean and movement is easy; if they have problems, the gullies are choked with rock and you have to crawl along, dragging your equipment or pushing it ahead of you to reach the face. If your places are new, then you might only have to travel in this way for 20 or 30 metres, if not…anything up to 120 metres is possible.

Today is one of those days and you have to crawl and climb all the way to the face. When you arrive there, the scene is alive with vigour and activity. The face is a wall of rock stretching away to your left and right. The one panel above you and the other below you. The hanging wall is smooth and undulating because the ground conditions are good and the rock is competent. The floor is a mass of rubble left from the previous blast. The cleaning has been poor. Only the few centimetres in front of the face are clean because the pikanins have washed it for the daily examination.

This examination is required by law in order to find any explosives that have perhaps misfired on the previous shift. The law states that you must do this but there is little chance of that, you have other things to do. About a metre back from the face, a line of sticks has been installed to support the roof. Arkey, the panel bossboy, is busy marking the holes to be drilled that day. Again, this is against the law. He is not allowed to mark holes; it is the work of the miner. He knows that you are there but he ignores you. He often marks the holes but has never done it in front of you before.

Today, everyone is late and, shift boss or no shift boss, the work must be done. You know that the miner is probably trying to make up for lost time somewhere else and cannot be everywhere at once. You are only too glad that the work is proceeding and you have no intention of throwing the book at anyone for putting in any effort they can to maintain production.

Once on the face, you look around. The timber crew is shovelling rock from the floor, clearing spaces about a metre square every 2 metres in a line from the bottom of the panel to the top. Today, they will install a complete line of matpacks. A matpack is timber support made by laying one mat of sticks crosswise on another until the roof is reached. This line of packs will support the roof permanently and is installed every time the face has advanced 2 metres.

Today, the face will be 2 metres from the support, tomorrow it will be 3 and the next day 4 metres. A new line of packs will be installed and so the unsupported distance will be back to 2 metres. One of the dangers of starting late is that this cycle is not maintained and chances are taken. The unsupported distance to the face becomes too large and the roof becomes unstable. Falls of ground occur, sometimes injuring or killing people, but always delaying that most impatient god of all—production.

It is satisfying to see that here the work is late but in hand. From the top of the working place, two sudden loud roars of a machine announce that Mafuta (fat one), one of the more prominent and aggressive machine boys, is growing impatient and wants to drill. A machine driller is proud and aloof, he considers himself a cut above the other workers. When he has drilled his quota of holes, he stops work and leaves the workplace. He does not demean himself by assisting in other work or by even remaining in the area.

Efforts on the part of management and miners to change this have met with little success. He puts his machine away, shoulders the two or three five-foot lengths of drill steel he has been using and returns to the waiting place. He will strip naked and shower under the nearest convenient tap. When he has dressed again in the clothes he took off early that morning, he will rinse out his working clothes, often not more than a G-string and flap of cloth, and hang them up for tomorrow. First to arrive at the shaft, he will be in the front of the queue for a cage and one of the first blacks to be on surface.

He handles his heavy machine with consummate ease, carrying it for kilometres balanced on his shoulder, or manhandling it in confined and awkward spaces with quick deft movements. He sneers at the efforts of management to get him to wear the earplugs designed to protect him against the continuous roar of the machine. Instead, he wears his deafness home to his family like a badge symbolising the elite band to which he belongs.

Arkey, the bossboy, shouts at him, "Tshia, tshia Mafuta! (Drill, drill, fat one!)"

The immediate roar of the big machine makes you want to shout too. Nothing signals that the job is going like the noise of the drilling machines. Suddenly, the tension lifts and you smile. "Vula Mafuta (Go Mafuta)," you shout.

Some of the timber crew smile as well. The bond between working men, black and white, ties you all together for an instant and the day might just be ok after all.

New Friends

Extract from the law:

...A student of the Government Miner's Training College or a student attending a training course in mining approved by the Government Mining Engineer, may be exempted from the age restriction of twenty years, and from such part of the qualifying period of experience for a provisional or a permanent blasting certificate, as the issuing authority may determine.

There were about twenty of us, mostly ex something or other. Ex-policemen, ex-railway workers or whatever. At eighteen, I was the youngest by at least four years and basically straight out of school. But I needed work quickly and the mining school asked no questions.

We had finished a two-week induction period and were to be transferred to another mine. We had to begin what was termed 'Basic Training'. The instructors had all taken great delight in telling us of the hardships that lay ahead. We would be doing the same work that the blacks did every day; shovelling rock, drilling the big machines, driving winches and installing the timber supports. This part of the training would take three months of the eighteen we faced before being qualified as miners.

We had been told many times that only six or seven of our intake would actually complete the eighteen months. Everyone knew that the bigger, rougher, Afrikaans men were considered more likely to 'make it', whilst Englishmen like myself and other more fragile people were all poor horses.

Two days previously, we had been taken to a working place and allowed to make up primers. A primer is a special charge, more sensitive than dynamite, which is used to ensure that the explosive packed into a blast hole ignites properly. To make a primer, you have to push a hole into the end of a small cartridge using a brass or aluminium spike. Into this hole, you insert the detonator of the fuse and then use a clip to hold it in place. Each drilled hole has to have one; so for this blast, we had to make up two hundred and fifty primers. What nobody told us, but afterwards quickly became obvious, was that this little exercise was to be our 'initiation'.

The initiation at this school traditionally took the form of an explosive headache. This is an intentional pun but is, nonetheless, an extremely accurate description of what goes on between your ears when it occurs. I have no understanding of the medical background to this phenomenon but know that initial exposure to the fumes of raw explosives results in a headache as devastating as the blast you are preparing for. The first sign that something is wrong is a slight nauseous discomfort. This is closely followed by a headache that develops into a pain of alarming proportions.

Nothing takes precedence over this pain; laughter, threats of physical violence, even ridicule are ignored to deal with the continuous existence of the mind-consuming beast. Your pulse rate climbs, you are ready to throw up and any movement or noise accentuates all of these discomforts. Before the task was complete, most of us were in serious trouble.

The humour of our affliction did not stop there. The two instructors who had perpetrated this inhumanity then led us to believe that by climbing the vertical ladders at the shaft to the next level, we would obtain some relief. They assured us that the fresh ventilation up there would soon blow away the effects of the fumes. They were well aware that any kind of physical effort would only increase the discomfort. So up we went. Some of us were vomiting before we reached the top of the long climb.

I managed to keep myself in check, and must admit, found that once resting in the fresh air, the nausea did subside somewhat. However, the headache persisted in its blinding intensity until I fell asleep that night. This episode caused us to view the coming basic training with some suspicion. Obviously, some of us would, as predicted, fall by the wayside. I was determined that I would not be one of them.

Although, I had been a fairly independent child and professed a certain hardness of heart as far as my family was concerned, the months prior to my enrolment at the Government Miner's Training College had been fraught with family trauma. My father was drinking heavily and the close-knit unit we had once been was splitting up. Two of my sisters had already left home and it was not a nice place to be. But, at the same time, I was very much in love with a young girl who lived nearby and not too happy at the prospect of being transferred somewhere else.

My life, it seemed, was taking a turn for the worse. But I could at least recognise a future of my own and to protect that future, I was resolute in my intentions to complete my training. The pressures of life, however, were just as set on countering this noble aim and the next few weeks became extremely difficult to bear.

Within a week, we were all transferred to a working mine and arrived together at the single quarters. The room I was allocated was no better or worse than any other room but when I opened the door, it seemed to me that there was a presence inside. Something that was hovering, waiting to attack any free spirit that might enter. The room was damp and the unpolished floorboards had a speckled appearance that spoke of neglect and disuse. Sagging against the opposite wall like an ashamed drunk was a mattress and beside it, lopsided because of a missing leg, its rejected partner, a metal bed frame.

I stepped inside and opened the cupboard; a space for a mirror but no mirror, a shelf and a rail above a dirty floor. Against the wall on my right was a washbasin, that was at least something. There was no curtain at the window, only a large gaping hole in the mosquito netting.

I walked over and looked out through the dirt into a parking lot. Slowly, I turned back to the room.

Shit, I thought. *I don't need this.*

Things improved by that evening. I made the room bearable and once eating, found that the food was good. After the evening meal, I phoned Elsabie and we came to the conclusion that I would survive. Later, I sat in my car watching the rain. I had just lit a cigarette when Randall Coetzee opened the passenger door. Blond and good-looking, Coetzee was one of the bigger Afrikaners and popular with the other trainees. He did not think much of me but was not beyond scrounging a smoke if he needed one. Other than that, he had been on my case from the beginning and never let an opportunity to needle me in front of the others go by.

"Got a smoke?" He asked.

I offered him the packet and he took two.

"Are you going anywhere tonight?" He asked.

"No," I said.

"Well, what about running us all into town? We can check out the talent."

I agreed, hating myself. I wanted to tell him that I did not like him, that I did not like his friends...but I did not want to be an outcast either and realised that for a few months at least, I was stuck with them whether I liked it or not.

The next day, we went underground to our new work situation. We went down one shaft, walked about a kilometre to another sub-vertical shaft, went down deeper still and then caught a train to the working place. This was known on the mine as the 'White Stope' and no blacks were allowed there. It was specifically designed so that we, as the future white supervisors, could physically experience all the tasks performed by the blacks on a mine. We met our two new instructors who seemed to be a little more human but also a lot tougher at the same time. They left us with few illusions about the future.

"You guys are here to learn what it is to be a Kaffir. For the next three months, you are kaffirs! You will load hoppers with shovels, you will lay tracks, drive loco's and drill holes with the machines."

He looked at me. "Some of you probably can't even pick up a fucking machine." Coetzee laughed out loud.

"Today, we will allocate jobs to each of you and go over the mining systems that we use on this mine. Tomorrow, you will go into the working places and start on your tasks. You will spend your days working, sweating and generally being mistreated. You and you," he pointed at Coetzee and Viljoen, "will be the bossboys and as such you will be responsible for the behaviour and output of the others."

He then divided us into two groups, putting myself and two or three others under Viljoen, clearing the blasted rock and some of the others, installing the pipes and tracks. The remainder of the class joined Coetzee and were to do the actual mining. I looked across at Coetzee and smiled, he did not appreciate why and turned to away to talk to someone else.

Life is not so bad, I thought.

In the ensuing weeks, I found myself teamed up with an older man, a Portuguese immigrant called Ferreira. I could speak very little Afrikaans and neither could he, so we tended to get put together on any job that did not require too much supervision. This was normally cleaning the ore from the gullies. We had to shovel the rock into a small truck called a 'ngolovan' and when it was full, push it about 80 metres before tipping it into an ore pass. Our quota was 8 tons each per day. Ferreira was strong and wiry, and I could not keep up with him at first, but by the end of the second week, we were working in unison. I was putting on weight and gaining some respect amongst the others.

One day when we were submitting our weekly report sheets, one of the instructors picked up that we had been loading well over our quota. Viljoen told him that I had been working with Ferreira on the loading. He looked at the figures and then at me and in front of everyone said, "You must be a lot fucking tougher than I thought."

That evening, Coetzee deliberately bumped into me in the dining hall.

"Aren't you the tough one!" He sneered.

I ignored him, so he turned to the others and said, "One day, Macarthur and I are going to sort things out, just for fun. Just to see how tough we really are."

He lifted his hand as if to hit me and when I flinched, carried the movement on as if he was brushing his hair. I was too slow, I ducked and everybody was laughing. Coetzee winked at the room and then pushed in front of me to get his food.

Later, Ferreira and I were playing cards and he said, "If Coetzee gives you any shit, you tell me, Okay."

I laughed. "I am a lover, not a fighter, this will not come to anything."

But Ferreira replied, "You are the hunted, and he is looking for a kill."

About this time, someone started smoking grass (marijuana) in my car during the night. Only two of us owned cars. I had sort of inherited mine from the family. It was old and lacked the odd mechanical niceties such as doors that locked or a starter motor. Consequently, although it was unlikely that anyone would steal it, there was no way that I could stop people from using it to sit in. I hated the smell of the grass that filled it every morning but had no idea of who was doing it.

One evening, I left my room to use the toilet and saw the glow of a cigarette in the front seat. I walked over and found Coetzee and Viljoen with another man whom I had never seen before. Coetzee got out of the car as I approached. "Hey, James, this my friend, Boetie." The stranger also climbed from the car and gave me his hand, saying something in Afrikaans. The smell of the grass was very strong and I was suddenly frightened. I decided not to say anything about the car.

Coetzee put his arm around me and I sensed that he was as high as a kite.

"Jimmy is my pal," he assured Boetie. "He is a fucking rooinek (redneck) but he is my pal. Soon he will be a proper boertjie (Afrikaner). Because why? Because I am going to teach him some manners, that is why."

Viljoen said nothing; he just sat in the back of the car with his head in his hands.

I turned and walked off towards the toilets. Coetzee called after me but Boetie distracted his attention and he turned back to the car. When I came past a few minutes later, all three of them were gone.

The next morning, the washroom was full of the news that Viljoen and another man had been arrested for possession of marijuana. We never saw Viljoen again. Coetzee came over to my table at breakfast and asked if I knew that Viljoen had been arrested.

"Yes, I heard," I said.

"Where did you go when you left us last night?" He asked.

"My room," I said.

He looked at me for a few seconds and said, "I hope so, McArthur. I fucking hope so. If you had anything to do with the cops coming here, you would be sorry. That's a promise." He turned and left the room.

"Shit!" Someone said. "The Englishman is dead meat." I ate my breakfast but did not enjoy it.

Later on, Ferreira came up to me. "Coetzee thinks you shopped them to the police last night."

"I wish that they had arrested him," I said.

He threw a quick glance at me, surprised. "He was not in his room when they raided it, so they went or to Viljoen's room and found only Viljoen and some other guy."

"Well, it was not me," I stated flatly. I told him what had happened the night before. "I was talking to them, they were in my car and Coetzee was bombed, but I left quickly. I don't trust those bastards when they have been smoking."

Ferreira shook his head and advised me to make sure that there was no grass hidden in the car.

Underground, we were told that one of Coetzee's group would replace Viljoen as bossboy and that I would join Coetzee's group to make up the numbers. Some of the other trainees looked at me and I knew they were wondering what was going to happen to me if Coetzee decided to give me a hard time. When I telephoned Elsabie after supper, I was as down as I could get. Talking to her helped but I did not tell her what was happening. Then she too, got in on the act.

"Do you remember that I spoke to you about a boy called David?" She asked.

"No," I said.

"Well, I used to like him a lot and next week when my mum and dad go to Pretoria for a few days, I will be going with them…" She hesitated.

"So?" I asked.

"Well, David will be there too and I will be able to see if I still like him…" She trailed off again. I could not believe what I was hearing.

"Elsabie, you can't. I mean, we are supposed to get married…" I stopped and took a deep breath.

"…Okay, you do that. Phone me when you get back and tell me whatever you want to," I said goodbye and hung up.

I was totally shattered. It was the early hours of the morning before I got to sleep.

Surprisingly, Coetzee left me alone for a while and after a couple of days, I almost believed that I was going to cheer up. But it did not last long and soon he realised that I was well and truly in his power. He started riding me again; in small ways at first and then more and more obviously. I would constantly tell myself that I did not belong with these people, that they were too rough for me, that I had not been brought up to live with this kind of behaviour. I almost convinced myself that I was somehow better than they were.

Deep down inside, I knew the truth; I would never be better than them unless I could play the same game with the same rules and win. I would not give up. So, I worked harder than anyone did. I worked more intensely than anyone and longer. I gave up smoking and volunteered for anything that was physical and unpleasant.

Elsabie came back from her experiment in relationships supposedly more in love with me than ever. She did not even notice David; she was missing me so much. So she said. I believed her. Then a bad ending to a miserable day made my mind up. Coetzee was in top form and had been on at me the whole shift. Despite some unexpected sympathy from some of the others, it was my battle and I decided that I was losing it.

On the way out of the mine, I experienced some South African mining culture that sickened me and I made up my mind to resign. We arrived at the main shaft to find the usual group of miners all waiting for the cage to come and pick them up.

Some miners are unaffected by the stress of their jobs but others are badly affected. I had noticed that nearly all miners behaved differently when they went down the mine to when they came out of the mine. When they go down the mine, they are often quiet. Not miserable, cheerful almost but not as rowdy as they are when they come out. At the end of the shift, they are brash and loud. They play cruel and obscene jokes on each other and on the blacks. On all shafts, there are those who live on the edge of their fear.

These people are unstable and because of the weight of their particular hell become so over reactive to shock as to be completely unable to control themselves. They are easy to recognise; they constantly flutter like birds. Their eyes are ever watchful and flicker from man to man, for they know no peace. Everyone loves to tease them or to have a laugh at their expense. They watch because they know that someone is going to surprise them with a shout or prod them or put a cracker in their pocket just to see what they will do.

One of the more common traits is the involuntary obedience to any sudden command. Today's victim was such a man. One of the miners moved up beside him in the crowd and suddenly shouted, "Kick, kick him!" Without hesitation, the victim kicked out at those around him. This was hilarious and everybody shrieked with laughter and waited for more. The victim, a tall thin man with 'FRIK' stencilled onto his helmet, swore and moved away to sit on one of the benches.

One of the others immediately went and sat down beside him and deliberately engaged him in a conversation about overtime. Once Frik was engrossed in this new interest, another miner sidled up behind him and undid the fly of his overalls. He took out his penis and held it forwards almost touching the back of Frik's neck.

"Hey, Frik," he called.

Frik, who was seated, turned and found himself peering at something in the miner's hand. He did not have time to find out what it was before someone else shouted out, "Suck it, suck it!"

Frik grabbed with both hands and sucked. Then, he stopped and stared, shocked at what he had in his mouth. He jumped back wiping his mouth viciously. "You fucking bastards," he yelled. He stood there gagging and spitting, wiping his mouth with both hands. "Fuck you, Billy, fuck you!" He kept saying, over and over again.

I looked around, Ferreira was laughing. Coetzee had virtually collapsed, and the instructors were laughing. All the way to surface in the cage, I kept thinking, *This is it; I am finished.*

On the way back to the single quarters, I told Ferreira, "On Monday, I am going to resign. I will stay on surface and go and see the principal and resign. Nobody will be in the office tomorrow because it is Saturday so I will go underground, but on Monday, I'm leaving."

He shrugged. "For me, this is a good job, but for you…maybe not so good."

On Saturdays, we always went underground an hour earlier so that we could come up earlier. My car would not start and eventually, rather than be late, Ferreira and I took the bus with the others.

We were at the shaft just after four o'clock and ready to go down at four-thirty. Ferreira asked me if I was still going to resign, and I told him that I would not be underground on Monday and then he would know.

On the train from the sub-vertical to the working place, I stretched out on one of the metal seats and dozed off. When we arrived at our crosscut, Coetzee leant over from behind me and banged my head against the side. "Wake up, 'Rooinek'."

"Oh, fuck off, Coetzee," I said.

There was a sudden silence. We all climbed off the train and Coetzee came straight up to me.

"What did you say to me?"

"I said leave me alone."

"No, you didn't, you told me to fuck off." He pushed me.

Ferreira stepped between us. He said, "Sort this out on surface."

One of the others took Coetzee's arm. "Leave the Englishman alone." Coetzee gave me a long hard look and turned and walked away.

Only one more day, I thought. But it was not to be.

An hour later, I was carrying sticks up into the working face. The sticks were mine poles about four feet long and three to four inches thick. As I climbed into the gully with some on my shoulder, someone pushed me hard from behind.

"Speed it up, McArthur."

I dropped the poles and turned; it was Coetzee. All my frustrations boiled sharply over and I swung at him. Hopelessly aimed and badly timed. I think I hit his shoulder. The only effect that it had was to bring him flying at me with both fists pounding. I fell over backwards and he fell with me, both of us grappling for some advantage. I found myself with his face locked under my arm and I squeezed as hard as I could, trying to get to my knees at the same time. He squirmed and heaved, but I hung on. I was angry and stronger than most people knew.

He could not breathe and was getting desperate. Then he bit me in the soft flesh under my armpit and the pain was too sudden and too intense. I let go of him and we both fell again. We were on our feet together and I tried to go backwards but could not. I was hard up against the side of the gully.

Coetzee picked up one of the poles I had been carrying. "I am going to fuck you up!" He said through his panting. "I am going to fucking kill you!"

He swung the pole from side to side in front of him. Then he lifted it, swinging from low down on the ground up at my face. I turned sideways and put my hands in front of my head. The pole hit my left hand against the rocks above me and my arm went numb.

I heard someone shout, "Stop, Coetzee! For fuck's sake, stop!"

Two of the others appeared from nowhere and then everyone was all over us. Someone took hold of my arm. "Shit!" He said. "Look at his finger. Oh shit, Coetzee, you've cut off his finger."

I looked down at my hand, my finger was not cut off, it was there all right but from the first knuckle to the top was a splintered mess. I turned my hand over: the nail was sticking out through the bottom of my finger. I felt sick and weak at the knees. Suddenly, it was terribly hot and I sat down. The next few minutes were confusing and difficult to remember. Coetzee kept coming over to me and saying, "I am sorry, McArthur. I am sorry." Then he asked, "Are you going to report me?"

Ferreira was busy bandaging my hand when one of the instructors arrived.

"What happened?" He asked.

No one spoke and eventually Ferreira said, "Coetzee and McArthur were transporting poles and Coetzee dropped one from up there and it fell on Macarthur's hand."

The instructor looked around and then at me. I looked at Coetzee; my finger was throbbing painfully. Everyone was silent again.

The instructor got angry. "What the fuck is going on here?" He asked.

"I could not get out of the way in time," I said.

"How bad is it?" He asked. "You bastards are going to fuck up my safety record."

"It's not so bad," I replied. "I think I have lost the nail." I immediately had a picture of what it had looked like and felt sick.

The instructor looked at me and asked if I thought I could make it out of the mine on my own. I assured him that I could and he put a field dressing on my finger. I got to my feet and set off to the station. As I moved down the gully, he called after me, "You just make sure that you are at work on Monday. I can't afford to lose any shifts to a reportable accident."

Later, when he found out that my finger was listed as a traumatic amputation, he read the others the riot act. All I wanted to do was get out of the mine and never see it again.

When I left the working places, it was about seven o'clock. There were no man carriages running so I had to walk. That walk started 2000 metres underground about 5 kilometres in from the main shaft. It went along a never-ending tunnel and up a shaft, then along more tunnels and up the main shaft. It went from one person to another and eventually, ended in a doctor's consulting room in the early afternoon.

By nine-thirty, I was in the dressing station on surface.

"I want to report an accident to my finger."

"Hau! The boss is not here. Is it a bad accident?"

"Yes, it is a bad accident." I hold up my bandaged finger, which now throbs without ceasing and feels the size of a lemon.

"Wait, I will put on a new bandage for you and then you can go to the hospital."

"Where is the hospital?" I ask.

"Just down the road, there behind the trees." Refusing the new bandage, I set off down the road.

Ten-fifteen—and about four kilometres 'just down the road', I arrive at the Duff Scott Memorial Hospital.

"I want to report an accident to my finger."

"Yes, you must wait here please." A pause. "You can sit down if you like."

Eleven-fifteen—A woman arrives. She asks, "What are you doing here?"

I say, "I had an accident and hurt my finger." I feel sick and shaky and decide not to stand up.

"This hospital is not for whites," she says. "Don't you have a private doctor?" After a long and difficult conversation, we both agree that I probably do have a private doctor and she draws me a map showing where the only two possible options have their rooms.

Twelve-forty-five—After walking up and down, asking directions from all and sundry, taking periodic rest breaks at bus stops or against the odd wall, I finally drag my finger, now a huge throbbing animal, into someone's consulting room. I am at the end of my strength. My finger is a thick mass of dried blood and bandage and the rest of my body is sick and faint. The woman behind the desk in reception listens to the painful details and mercifully takes me into a cool room where I sit and rest my arm on an enamel table.

She leaves the room and returns with a tray of scissors and other assorted kit. She washes her hands and comments on the warm weather.

I say, "Please, just look at my finger."

We get more professional and the agony starts. As she begins to reveal what lies beneath the bandage and it becomes apparent that half of my finger is more a part of the bandage than it is of me, she gives up and fetches the doctor.

"Oh my God," he says. "When did this happen?" I told him that it was about six-thirty that morning. He approaches me with a syringe and injects the base of my finger with some clear fluid and then stands back and waits…

"What is he waiting for?" I ask myself, but when the pain starts to fade and my finger becomes a dead lump instead of a vindictive one. I understand.

This doctor and his receptionist are wonderful people. The relief is fantastic, and I become cheerful and friendly. I smile at the receptionist; the warm weather is nice after all.

I did not resign on Monday; I was in the hospital waiting for them to amputate the mess at the end of my finger. But Coetzee did. Ferreira told me afterwards that the instructors said that they had lost a star miner. I never went back to the white stope for the rest of my basic training. Instead, I was on light duty for several months. My finger went through a whole series of problems; trouble with stitches, infection and eventually, a second operation.

As a result, I spent most of my training in the stores, geology department and survey office. This did my image an immense amount of good and most of my fellow students were sure that I would be a manager one day.

I received R170.00 as compensation for the loss of my finger and spent it all on a wall clock for Elsabie. What is left of my finger still works but the wall clock doesn't. One day, I will have it repaired and send the bill to Coetzee.

The First Time

Extract from the law:

The manager shall further…
…not permit any incompetent or inexperienced workman to be employed on dangerous work or on work upon which the safety of persons depends…
…In no case allow the ganger to have charge of workmen scattered over more places than can be inspected without undue exertion within forty minutes.
…not allow any ganger or miner to be placed in charge of a gang or gangs of workmen which, regard being had to the number of persons therein or to the nature or position of their working places such ganger or miner is unable to supervise efficiently in accordance with the requirements of these regulations…

Every morning at seven-thirty, I met my future wife at her garden gate and walked with her to her bus stop. She was seventeen and on her way to school. I was nineteen and on my way home from work. She would learn and I would sleep. She was a creature of the day and I was one of the night.

In the late evening of the previous night, I had arrived at the shaft of the mine on which I worked and entered the change house. My colleagues and I had exchanged greetings and then fallen silent as we donned our working clothes before going underground. Each change house has an atmosphere of its own. This one was a huge draughty hall with rows of countless baskets hung from a steel web high above our heads. The baskets were allocated, one to each individual, and used to keep personal belongings secure whilst you were on shift.

I unlocked the chain on which my basket hung and lowered it until it was waist-high. I sat down on the bench and sipped a cup of tea and looked at the men around me; each getting dressed in a silence broken only by the occasional greeting. There is something strange about going underground at night; the men around you lack the spirit that fuels the bustle and activity of the day shift. Men arriving for nightshift seem to be somehow depleted by the day without perhaps having contributed anything to it.

When it was time for our shift to go down the mine, we came out into the cold night air and walked across the concrete apron carrying our bags to the shaft. Each of us booted, helmeted, and kitted in our own style but all with a uniform that said, "I am part of this brotherhood of miners. I am one of these men that go down the mine to earn the money that lesser men have no right to."

Some of us affected traditional clothing; thick overalls gathered and tied at the knee over army-style leather boots. Other younger men, myself included, preferred calf-high rubber gumboots with thick socks folded down over the tops, accentuating our calf muscles. We sneered at overalls and wore tight boxer shorts with no shirts, condescending to throw our leather mining jackets across our shoulders when we walked across to the shaft head during the winter months.

Helmets, boots and belts were one's claim to individuality and one tried to make sure that they reflected years of experience and a wealth of wisdom about how to survive in a world of heat, sweat and constant contact with death.

So, on that night, we stood, as on most nights, each alone with his private thoughts in the crisp, late night air and waited for the cage that would drop us 2,500 metres into the bowels of the earth. 2,500 metres into a world which exists only in the memories of men, for no one on surface reflects on paper the truth about what goes on underground. This was something I was to learn brutally and finally in the next few hours.

You cannot allow yourself to be soft or sensitive if you are trying to make your way in a tough world. A cloak of iron is required. Sensitivities must be buried and their existence denied. It happens occasionally that fists must be clenched and blows struck in order to demonstrate that your make-up does not include such weaknesses. But if they are there, they are there and rise up at awkward times to embarrass you. On this night, they were to rise up and embarrass me but were struck such a blow that they have never again surfaced with the same assertiveness or confidence.

At about twelve-fifteen am, I had toured most of the working places and the work at all of them was proceeding without any signs of problems that were not identical to any other night of the week. I had not yet qualified as a miner and so was not technically allowed by law to supervise the workplaces. However, I knew that I was as capable as any one of the old 'has-beens' that normally ended up working the nightshift.

This in spite of the fact that I had not even had the amount of training my fellow learner miners had received. I had lost the tip of my index finger in a fight and had spent seven of my twelve months training, not underground but on surface, in administration work.

I was not too concerned about the way the work was proceeding and walked out to the main haulage to wait for the shift boss. The area where I worked was about 2 kilometres from the shaft and it was common for some of us to hitch a ride on one of the ore trains although this was strictly forbidden. Sure enough, when the next train came in from the shaft, rocking from side to side along the tracks, it slowed and my shift boss jumped from the last hopper. He sat down beside me and I offered him a smoke.

"What does it look like inside?" He asked.

"Everything is fine," I replied and read to him from my notebook a blow-by-blow description of each of the working places and its problems.

"Have you been there yourself?" He asked, "or did you get all that from the bossboys?"

"No, I went into each place myself. The only place I have not been to yet is 26A."

He was angry and swore violently.

I felt my hackles rising. He could not talk, he never went into any of the working places himself and was an arsehole anyway.

"I am going there now," I said, "and I have told all the workers that they will find me there if they need me. That is the worst place, that is where I plan to spend the night."

"Well, I hope for your sake and mine that nothing goes wrong. Your miner knows that he should not force me into giving him a night off when the places are so dangerous. He doesn't give a shit; he only cares about that car of his. If anyone finds out you are working the places alone without a miner in charge, I will be for the high jump. The mine captain will chew my balls off."

"You should have made him take a day's leave," I said. "That way you would have been covered."

"Ag, you do not understand." He stood up as a train approached. "I have other problems, so I won't visit the places tonight. If you have lied to me, you will be in front of the mine captain tomorrow, I want you to get off your butt and go to 26A before someone gets killed."

He waved the train down and climbed up onto the hopper.

"You don't come out of the mine unless everything is fucking clean!" He shouted as the train pulled away.

I sat on a rock and ate a sandwich, washing it down with some iced water from my bottle. As I watched the cockroaches scurrying over the rocks in search of the food they knew was there somewhere, I wondered about how stupid the management was. Surely they knew that the nightshift crews did more or less as they liked, that men went underground drunk. That they sometimes signed on and off on the timesheets and never went underground at all. Just like the miner whose working places I was looking after had done tonight.

Perhaps they did not want to know? Mentally I shrugged. I did not care, the miner in question had given me R20.00, partly to look after his places and partly to keep my mouth shut. Working underground meant breaking the law and that was that; breaking six laws in a night instead of five, well so what?

I set off to walk to 26A where I planned to spend the rest of the evening reading my book.

26A was a raise connection with three working faces. They were old places and far in from the centre gully. The crew that cleaned there was small; with only ten workers in total. They worked under the supervision of a bossboy called Simon. Simon was about thirty-five years old, short, very tough and intelligent. He was always active and aside from a constant suggestion of liquor on his breath, fairly reliable.

The night before, he had quizzed me on my background and when he found out that my parents lived in Capetown told me that he had heard that blacks could get contract labour on the fishing trawlers there. He explained to me that his insides had told him for many weeks that the mines were a bad place for him and that he should get out as soon as he could. The story did not mean much to me then; the blacks were always telling me that the mines were a bad place.

"Why do you young white people choose to work in the mine?" He asked.

"For the money," was my reply.

Simon and I went into the first place together. The winch drivers were busy pulling the ore down the gully and we had to wait for them to stop before we could continue.

Within ten minutes, the cable snapped, work stopped, and we could crawl over the rocks towards the face about 60 metres away. As we went in, we met the face winch driver on his way out to help the gully driver pull the cables together.

"What is it like inside?" I asked him.

"Here where I am working, it is ok," he replied. "But up there," he pointed at the next face. "It is very bad; big rocks that must be blasted."

I looked at Simon, who shrugged. "It is always bad there," he said. "I blasted some big rocks earlier and I will have to blast again."

"I don't like you blasting, Simon, you are not allowed to blast."

Simon laughed and shook his head. "You just leave the job to me. I was blasting big rocks when you were still a baby."

He was right, most of the bossboys in the mine ran the jobs for their miners. They were in fact the backbone of the nightshift. They had years of experience and watched the white nightshift cleaners come and go whilst month after month, year after year, they cleaned more tons of rock from the working places than half of the white miners would ever see.

I stopped and sat down on the dirt. "Simon," I asked, "what is the place down here like?"

"It is fine," he replied.

"And that place up there?" I pointed towards the place above us.

"That is bad." He shook his head again. "I do not know if we can get it clean by the end of the shift."

"Ok, let's give this place a miss and go to the bad one. We can go up through the back here and look at it."

We got up and left, cutting up through the worked-out area above us to the next gully. It was a disaster area; the day's blast had obviously brought down a major rock fall, knocking out two of the supporting timber packs on the side of the gully. Towards the back it seemed better, the huge slabs supported by large packs. To the front, the winch cables and other rigging equipment were lying under huge blocks of stone, some of them more than a metre thick. The roof above us was huge and open, the missing packs exposing a large expanse of unsupported rock.

It took Simon and myself just over an hour and several blasts to clean a path for the scrapers to operate. Finally, the cables were rigged and the winches ready for pulling. I left Simon there. I told him to make sure the job was going well and then to come and report to me at the miner's base at the top of the raise connection.

I visited the third working place and found that the crew had nearly finished cleaning and would be leaving for the shaft soon. Ten minutes later, I was sitting at the miner's base, or box, as it was known, in cool ventilation and busy with my book and the remainder of my sandwiches.

All working areas underground have this unofficial office known as the miner's box. Here, the responsible miner keeps his tools, his lunch and hangs his jacket. On the nightshift, we choose the most comfortable of them and use it as a base. Because a nightshift cleaner cleans for two or three dayshift miners, his area of responsibility is very often too large to cover more than once in a shift, if it can be covered at all. This means that it is convenient to have a central reporting point where all the bossboys can find you. When they have completed the work, or have a problem, they know where you will be.

The nightshift cleaner waits there until the last one has reported and then leaves the mine. Most of the better cleaners make a point of visiting each working place at least once in a shift, but to comply with the law, one is supposed to be with each crew when it enters the working place. Normally, the cleaner will go to the panel where there is a known problem and if the other crews had to wait for him to arrive at their place, they would never get started.

The idea of sticking to the law and then having to explain to the mine captain why the faces were not clean for the next day's blast was not to be contemplated. At best, you would lose the cleaning contract and your all-precious bonus. At worst, you would get punched into the middle of next week and then lose the contract.

So, I sat and waited, reading the western I had brought with me. Two-thirty am dragged on to three am and was slowly inching its way towards three-fifteen when I saw a light approaching from Simon's raise connection. When the man arrived, I saw he was one of the drivers from the upper panel. He was sweating and panic-stricken.

"Boss, boss," he shouted as he approached. "The bossboy is under the rocks, he is under the rocks! The mine has fallen."

I grabbed my belt and lamp and followed him as he turned and went back the way he had come. My heart was pounding and I could feel a black despair in the base of my stomach. This was it; this was what I had heard discussed, laughed about or even ridiculed in a number of change house bullshit sessions. What would it be like? I had never seen a dead man in my life. Now I would have broken bodies and torn flesh to deal with.

"Oh God," I prayed as we climbed down the various ladders and tunnels to 26A. "Please don't let him be dead, please don't let him be dead."

When we arrived, all the boys who worked in 26A were already there. Moses, one of the drivers, was pulling on a scraper rope as we emerged into the fallen area.

"Where is Simon?" I asked.

He pointed to the edge of the working place, an area we call a south siding. Timber lay in all directions, some of it protruding from under a solid slab of rock at least 6 metres long and 3 metres wide. Along its edges were more pieces of broken rock ranging from small to almost as large. I started forward but Moses grabbed my arm as another slab broke free from the roof and fell, its crash as it hit, sent choking dust in bellows down the gully. I stopped and looked around at him, noticing for the first time that he was actually busy with the cables.

"What are you going to do?" I asked.

"I want to tie this cable to the big rock and pull it off Simon. The rock is too big for us to lift."

I had visions of Simon being ground to pieces as the rock was pulled off him.

"No!" I said. "No! We must lift that rock carefully. We have to lift it."

He looked at me and said contemptuously, "Lo muntu, Yena fiele (The blackman is dead)."

"I don't care!" I shouted. "We must try to lift it."

He shrugged as they all do when faced with a white man's refusal to listen to reason and then turned away. I looked around; all of them had stopped when I shouted at Moses. All of them stood waiting like deactivated robots waiting for some command to set them in motion. The shock settled slowly on us and I realised that it was I who must push the buttons. But I did not know what buttons to push. Simon had to be alive, it was inconceivable that he could be dead. The rock had to be lifted.

"Moses, you take four people down to the crosscut and get some big sticks. We must support the roof here where we must work."

He looked up at the roof and then nodded.

"Then you look for the first aid boy and tell him to come here with a stretcher and his equipment. Go! Go now."

I pushed him and motioned some of the others to follow him.

I called to one of those who remained behind, "You, you take two people to help you and bring a piece of four inch pipe from the pipe bay. Bring a long piece, bring a new piece."

I looked around at the two boys left with me. "Get a shovel and some pinch bars."

They left and suddenly, I was alone.

I looked at the rock under which Simon lay and started to move out into the area towards it. The roof cracked above me and some dust settled slowly through the air to my left. I stopped and waited without moving. I did not look up. Nothing happened and I moved forward again, being very careful to touch as little as possible. I could see no sign of Simon. I bent down and picked up a rock and threw it to one side. I moved another one and then another.

After some time, one of the boys returned and unquestioningly moved out into the open area to help me. Together, we lifted a larger one and moved it. He stood and looked around him.

"Boss, we must move the rocks over there. Simon is between the packs."

He showed me where the remains of two packs protruded from the rubble. I nodded and felt the tears swelling in my throat. Shit! I could not do this. I could not dig a dead body out of the rock, this was not what I had been paid R20.00 for. Simon was not dead; this could not be happening.

But there was no one else and if I walked away, then Simon was definitely dead. I tried to clamp a lid down on my feelings but had to turn away to hide the tears, so started lifting more rocks and moving them out of the way.

Soon more of the boys returned and we prised, lifted and shovelled rocks, sticks and debris for an eternity. Slowly, we excavated a hole in the ore, tunnelling down beside the rock that lay on Simon.

Moses returned with some sticks and fitted them as best he could on top of the larger slabs of rock to try, in some measure, to support the roof above us. He looked at me and said, "The roof is talking."

I listened. It was; little whispers of sound and movement interspersed with sharp cracks and subdued pistol-like shots. I watched a small cloud of dust settle slowly in the still hot air. The ventilation was non-existent and we were all suffering, sweat running grey rivulets down dust-caked faces.

What would the mine captain say if the roof fell again and crushed more of us? What if instead of one dead person, there were three or four? Well, I would probably be one of them. Surely, they would understand that I had to try? What if Simon was still alive?

"I can see him!" One of the boys shouted. "I see him!"

Roughly, I pulled him out of the hole we had dug and climbed down to look. It was difficult to get low enough to see anything. I took my helmet off and held it so that the beam of the lamp shone under the rock. Yes, I could see him too; at least I could see his boot and some of his leg.

"Moses, bring the pipe and we will try to lever the rock up and drag him out."

All that had been running through my mind since I had first seen the rock was some expression or other about "Give me a long enough lever and I will move the world."

Moses knew it was hopeless before we even started and so did some of the others. But I was the white boss so they said nothing and we tried anyway. We pushed the pipe down the slot we had dug and forced a large boulder beneath it at the fulcrum.

When everyone was positioned as I wanted, we heaved on the pipe with no result.

"Pull!" I shouted. "When I say pull, we must all pull together…Now! Pull!"

Muscles and will strained together and nothing moved. I closed my eyes and swung my full weight on the pipe. "Move, you bastard, move!"

"Pull, pull!" I called again and again.

Suddenly, Moses let go. "Aikona baas! It won't lift. We must pull the rock with the scraper rope."

"I say you will not pull this rock off with the scraper. I will never agree. You will kill him. Maybe he is still alive under there. You can fuck off! Fuck off out of here! Do you hear me!"

I was angry and frightened and yelling at the top of my voice.

Moses looked at me and shook his head. "He is dead, baas, Simon is dead."

For some seconds, there was silence and no one would look at anyone else. Then someone said, "Maybe we can jack the rock up with a hydraulic prop."

It was a brilliant idea and I leapt at it. "Moses, bring a prop, you and you, go and fetch the pump machine."

Once again, there was hope and this time, it was real. The boys turned and went to fetch the equipment.

As they left, I saw lights crawling along the gully towards us. The first aid boy had arrived with an assistant. He said little except to ask where the victim was and then told me he was not prepared to work where it was not safe and that we would have to remove the body to a safer area before he could examine it.

I ignored him and when the others arrived with the prop, we wedged it into the hole with its top under the edge of the rock. Moving this rock had become all that mattered now. Simon had somehow receded into the background. This malevolent piece of grey quartz was mocking us and had to be lifted, not rolled or slid, but lifted. Clean and clear, straight up into the air, and that is just what we did.

Eventually, at about five-fifteen am, with the hydraulic pump snorting and wheezing in the background, we slowly but surely forced several tons of rock into the air. Tipping it along its far edge until there was at least a metre clear where we were working. Moses and some others forced rocks into the gap to wedge it and for the first time, we could truly see Simon.

I suppose your first fatal is always the worst and although, I have seen more death since in the mine, nothing has remained as vivid as this. He lay, as if asleep, on his side. His knees were curled up into his stomach, his back towards us. He had been wearing his overall wrapped around his waist leaving his back and shoulders bare. The skin of his back was split open from his shoulder blades to his belt, the fat, yellow and flecked with red, some white pieces of bone protruding from it.

I turned away and sat down, my legs weak. I felt hot and tired and bowed my head between my legs. The bile rose up in my throat but I would not allow myself to be sick.

Some minutes later, Moses touched me on the shoulder and I looked up. Simon was a blanket, rolled up and placed on a stretcher.

"He is dead," Moses said.

I nodded. "Is he in one piece?" I asked.

The first aid boy answered, "No, he is not. His hand is missing and I cannot leave without his hand. His right hand is still under there." He waved at the gap under the rock.

I stood up but Moses put his hand out. "It is enough, baas, it is too dangerous and it is late. We must go now."

I pushed him aside and crawled under the rock, holding my lamp in my hand. I played the beam around the confines of the space where Simon had lain. Over in the far corner, I could see something but I could not reach it. I crawled backwards out of the hole. Everybody was watching me. They were tired and covered in dust and sweat, robots once more.

"It's too far," I said.

The first aid boy shook his head. "I will not go without it," he replied.

Moses swore at him and told the others to pick up the stretcher. They waited for me to say yes.

I put the cap lamp down and took off my belt and my battery, took off my helmet and crawled back under the rock. Without the lamp, I could see nothing, it was pitch-black. I could only feel and smell. The rocks underneath me were sharp and pressed painfully into me the further I forced my way in; the dust filled my nose and I was about as far in as I could get. I had my head wedged against a smooth surface above, against dirt and grit below. I pushed my arm ahead of me and groped this way and that until at last I felt my fingers touch something that could only be another finger.

Closing my mind to the feel of the thing. I gripped it and pulled. When it came free, I reversed out of the hole.

I walked over to the first aid boy and dropped Simon's hand onto the stretcher. "There's your fucking hand!" I said.

Of the long journey from 26A to the shaft, I remember little except the stream of dayshift workers coming down the tunnel in the opposite direction. They laughed and shouted, jostling each other and fooling around until they saw the stretcher with the blanket over the head of its occupant. Without exception, they would look at the stretcher, take note of the blanket, fall silent and then raise their eyes to me.

No matter that I was torn and covered in filth. No matter that it was me that had preserved whatever dignity Simon had left when God's hand smashed him down; it was to me that they raised hurt, accusing eyes. The white man had invented this hole in the devil's kingdom, this place of humiliation and death. It was the white man who must answer for it.

In all my years on the mine, this knowledge remained with me. The industry has brought prosperity and hope to millions of blacks. It has brought about a blossoming of civilisation and a widening of horizons. It has financed careers, schools and hospitals, built homes and roads.

It does not matter; the simple mind looks for the immediate source of any pain and the mines have been at the forefront of much grief.

I wanted to scream out loud, "I did not kill him, nobody killed him. It's not even my working place, his miner took a day off and left me there."

But in truth, I did kill him. I was too cocksure. I willingly accepted R20.00 to aid a scheme to delude the management. I stood up and said to myself and therefore everybody else, I can run this job. Another more experienced person might have called the boys out, or even just stayed until the job was finished. Perhaps I could have examined the area more carefully. Obviously, if someone is going to die, then you can never be too careful, but when is someone going to die?

The news travelled ahead of me and when I got to surface, the mine captain was waiting on the bank. Several of the dayshift were standing around and wanted to know what had happened but before I could say much the mine captain took me to his office.

He walked across to the window and with his back to me, asked, "Where is your shift boss?"

"I don't know," I said.

"Where is your miner?"

"He had the night off."

I could see his neck start to go red and knew that trouble was coming. He turned and looked at me. I realised that he was angry but also frightened.

"Your shift boss was out of the mine and off home by one-thirty this morning. Your miner was not even underground! Who dug that fucking kaffir out of the rocks?"

"I did," I said.

"And who was in charge of the working places?"

"I was," I said.

"You!" He shouted. "You! You are in charge of nothing. Nothing, do you hear? You are a fucking learner! You are still wet behind the ears, you little shit!"

I shook my head in bewilderment.

"How much did the miner pay you? You think I don't know what goes on? You think that I never worked on nightshift before?"

Then why don't you stop it? I thought it, but did not say it.

"Go and shower, say nothing to anybody or I will personally tear your miserable little heart out. When you are finished, come back here."

He left the office behind me and as I went into the change house, I saw him enter the manager's office.

That is strange, I thought. *It's not yet seven o'clock, what is the manager doing here?*

When I returned to the office, he was back and calmer. He told me to sit down, and said, "You did well to get that boy out, but there are problems here that are beyond you. Do you understand that?"

I nodded and he continued.

"I have spoken to the manager and your instructor at the school and we have arranged for you to have a few days off. You can go home, go away if you can but do not show your face around here, and do not…repeat, do not discuss this with anyone. When you come back, you will be transferred to another shaft."

"I don't want…" I hesitated.

"You will just do what you are told. Do you want to face an enquiry? Do you want to go to prison? They put people like you in prison, you and your miner and your shift boss. You go and spend whatever money they fucking gave you to work those places and don't talk shit to me."

He caught himself and stopped, then said, "Look, I know you are not entirely to blame, but on the other hand, you have been party to something criminal and someone has died because of it."

"Wait," I said. "I did not do anything that everyone else doesn't do. All the learner miners look after working places. Some of the miners even pay them a share of the bonus."

I got a little bolder but could feel myself shaking and knew that there was a tremble in my voice.

"No one can see all those working places in and still get the work done. The bossboys see the crews in by themselves, they blast big rocks, they break the law all the time. You all break the law!"

At first, I thought he was going to get angry but then he leant back in his chair and frowned. "I don't know that this is happening, the manager does not know that this is happening, my shift bosses never tell me that this happens. Perhaps you are not man enough to do the same as everybody else. Perhaps you should not be working in the mine at all."

"They all know," I said.

"Who knows? You bring one person who is prepared to admit that they cannot handle their contract, as you say they can't, and I will believe you."

He looked at me straight in the eyes. "You just do what you are told and in a week's time, everything will be ok. The mines have been rolling along for quite some time without you and you will change nothing. You must make up your mind as to whether you can roll along without the mines."

I just made it in time to meet Elsabie at seven-thirty am as usual.

"What happened to you?" She asked. "You look as white as a sheet."

I explained that I had dug a dead black out from under the rocks.

"The mine captain is a bastard and the mine sucks," I said.

She put her arm around my waist and we walked down the road.

"He gave me three days off," I said. "To recover," I added.

"Well, he can't be all that bad," she replied.

"No, I suppose not," I said.

Malawi

Extract from the law:

Detonators:

Instantaneous electric detonator made with a copper capsule and of a strength not less than 60 as defined in the regulations framed under the Explosives Act 1956.

It does not matter here, where I work, what my name is, because everyone calls me Malawi. They call all of us Malawi. I came here with about two hundred of my brothers; we are all young people and have good happy faces. Everyone recognises us straight away. 'Hey, Malawi, come here!', 'Hey, Malawi, do this!'.

I don't mind this because Malawi is a good country, and I am proud of it and our leader. He is known here in South Africa as the doctor and nearly all the white men know about him. He used to work on the mine as a tip boy just like I do. Perhaps I will also lead a country one day. But first, I must learn to speak English.

When I arrived at this mine, I received good strong clothes and other things that I need for my job like boots and a helmet. I was taught many things that I knew about already; like how to keep clean, and how to cross the road. But they did not teach me how to speak English; instead, they taught me to speak Fanakalo, which is the language of the mine. I want very much to learn to speak English.

In my room, which I share with thirteen other Malawis, we discussed this amongst ourselves. The others tell me that when we work for an English miner, he will teach us English. Only he must be an Englishman. You must not ask an Afrikaner to teach you English, they will get cross. The blacks on the mine call the Afrikaners 'Boere'. They also say that the 'Boere' are strong and aggressive people who are sometimes fair but always ready with their fists and who hate the English.

I have made a friend who eats with me in the dining hall. He works for a 'Boer' who is the best miner on the shaft. Everyone who works for him gets a small share of his bonus. He does not have to do this but he says that his boys earn the bonus so they must get some of it. If he does very well and makes his target, then he gives his team a party. He buys meat and beer and they all get drunk.

I asked my friend if he thought that this man would teach someone English.

My friend laughed. He said no, his boss would never teach anybody English.

Sometimes the day is hard on this mine. I am told that it is the same on all the mines. I get up at half past three and I eat a lot of food before I go to the shaft. I need this food because I work hard and I get tired if I do not eat. We fetch our lamps from long racks and then stand and wait to go down the mine. If you do not get the early cages, bad things can happen to you. Sometimes on the later cages, you must get in the bottom half and the white miners get in at the top.

They always play jokes on each other but sometimes on us as well. Once they pissed through the floor onto us. Most of us thought it was funny and it was because I laughed too, but it didn't fall on me.

We always get to work before the whites arrive and are already busy, working hard before they even see the workplaces. Some of them are strong and some are lazy but they are all clever; they know many things. Always if you have a problem that stops the job, that you cannot fix, then you must think quickly because if they come then they just say 'Do this, do that!', and everything goes again. Then they call you stupid. It is not nice to be stupid. It is very important to be clever because then they give you a better job and you can become anything you want. The doctor, our president, in Malawi, used to be a tip boy on the mines and he is now the leader of our country.

I am very happy because I have been transferred to a job on nightshift. I was a tip boy on the dayshift working for a 'Boer' but now I am working for an Englishman and he is a good man. He asked me my real name and what village I come from and then he told me what I must do to make him happy. I think he will teach me to speak English.

The others told me that they call my new boss 'Doppie' because, although he never hits anyone, he gets very cross very quickly. He suddenly explodes without warning and then everyone is frightened. But he is really called 'Doppie' because even if he does explode, he does not do much damage. But I do not want him to explode to me so I keep my tip clean and look after it well. My bossboy,

Elijah, says that I am working well and the other workers are helping me a lot. I am learning a lot about the mine.

Last night, Doppie said that I was doing well. I was very proud and told my friend in the dining hall that I never wanted to work for anyone else. My friend has worked for his miner for twelve years. He knows that when he goes home for a holiday and then comes back, his boss will make sure that he goes straight back to his old job. I would like to know that I would always go back to my friends on nightshift.

Doppie says that he will teach me to speak English. That is wonderful. I know he will be a good teacher; he is a very clever man and sometimes reads books underground with very small letters and many, many pages. He said we must start straight away and taught me to say, "I am a stupid tip boy."

He says that when I have learnt to say it properly, then he will tell me what it means and then teach me something else. I am going to keep this a secret until I can say it perfectly, then I will say it to my brothers in the room. They will be jealous because I am advancing myself.

Last night, Doppie came past the tip and asked me, "Hey Malawi, what are you?"

"I am a stupid tip boy," I replied.

He laughed and walked on. "That's perfect, Malawi. That's perfect."

I am very ashamed and today, I went to the mine captain and asked to work for someone else. When I told the others in my room that I could speak some English perfectly, they were amazed, but when I told them, "I am a stupid tip boy," they laughed and told me what it meant. I did not believe them, so I asked my friend in the dining hall.

Whilst I was waiting in the office to speak to the mine captain, Doppie came in and asked me, "Hey, Malawi, what are you?"

I did not answer him. Tonight, I will be working in a new section and I will never work for him again. When I learn to speak English, I won't speak to him.

Mud Rush

Extract from the law:

The manager shall take reasonable precautions to ensure that every person employed in the workings of a mine is safeguarded against inundation by water or mud or a flow of rock, sand, silt or other material.

James had worked for me for about two months before it happened. He and I were different from most of the other young miners on the shaft. We both had a greater appreciation of life's possibilities than the others did. I was a shift boss at the time and like James, at the start of my mining career. I matriculated and became an official; he had left school early and become a miner. Another aspect that we had in common was that we were both English and yet fairly straight. The other young English miners were a little strange and not really considered to be worth a great deal by either the management or their workmates.

Brian, for instance, was a wild bearded fellow who came to work on a skateboard and openly declared that he only worked six months of the year. Things often happened to Brian that confirmed the old timers' unspoken but obvious opinions. Once he decided to take over the work of the guard boy on one of the ore trains. Unfortunately, he fell asleep on the last hopper and the driver rode him into the far reaches of the mine, uncoupled the hopper and left him there. It was halfway through the next shift when he finally awoke and found himself with a six-kilometre hike back to the shaft.

We all thought it was as funny as hell, but one old Afrikaner only grunted and spat into the dirt. "What do the kaffirs think?" He asked.

The only other Englishman on nightshift was a short violent man who shaved his head bald, wore a ring in his ear and called himself 'Curly'. He was known as a dagga user and often laid off work. It was not really surprising that James and I found common ground.

Nightshift was not difficult for me. I liked not having the mine captains and managers breathing down my neck. On arrival at the shaft at about eleven pm, I usually checked the time sheets to see if my shift had gone down and then read any instructions that the dayshift had left for me in the nightshift intercom book.

I liked to read the instructions left for the other shift bosses as well; they were sometimes abrupt and well-spiced with obscenities. Not that I was entirely without fault, I had my share of written abuse too. Then I would change and go underground.

After the shift was over for the shift bosses, normally around four am, we returned to surface and slept in corners, on benches or on the floor of the change house until the dayshift started to arrive and pushed or shouted us awake. The miners came out of the mine at any time from five to seven am and would go straight home after showering. The unfortunate officials like myself had to report daily to the various mine captains in charge of our sections. James always crept into the official's change house before he went home and gave me an update on what was the state of his work.

"John, John," he would whisper. "Listen, all the boxes are empty except for 19 and 20 in 7B. It's that Y leg again. It is hung up. I blasted but it is still stuck. Get the dayshift to blast some more. We will need it tonight."

I would nod and he would slip away in the dark, his work over. I never had any trouble with him and we were friends; I suppose I should have trusted him.

I had two white nightshift cleaners working under me, both with fairly large sections. They cleaned the working places of the three top miners on the shaft and were considered to be good men. James was in charge of the ore trains and the removal of the ore on 62 level, which served the area where these men cleaned. His job was to make sure that whatever ore the cleaners tipped into the boxes at the top was removed and taken to the shaft as efficiently as possible.

As always in this kind of situation, there were not enough trains, the tracks were poor and derailments common. The ore passes were nearly always full, and it was all James could do to stay ahead of the tipping operations above him.

On this particular night, the dayshift shift boss had written in the intercom book:

John

Box 6 is full of stuff; it was reported to me by the dayshift when they came out. We did not blast that panel yesterday so we drilled long steel and there will be a shithouse full of stuff there.

McArthur must go there first and pull it empty!

I opened the cleaner communication book and saw that the dayshift miner, Peter Cameron, had written:

N/Shift 62 level 17C Pull boxes 6, 7 and 9. <u>PULL BOX 6 FIRST</u>

I walked to the shaft and phoned the first aid station on 62 Level and told the first aid boy to find James, the boys called him Doppie and tell him to phone me back straight away. About half an hour later, the change house bossboy called me to the phone and I spoke to James.

"Are you pulling box 6?" I asked.

"No," he replied. "Box 6 is not full. It's empty. I emptied it the night before last and when I checked it last night it was still empty."

"Well, the dayshift says its full," I said.

"I know, I read the book too, but my bossboy says that he got a message from the dayshift bossboy to say they have not tipped into box 6."

"For fuck's sake, James," I replied. "Who runs this bloody mine, us or the bossboys? Go and check the box yourself…and pull the fucking thing. OK?"

There was silence at the other end of the phone.

"I'll see you later," I said and hung up.

I went in on 59 Level and checked with Sarel Willemse that his section was ok and then came down the centre gully of the 18 line to 62 Level. When I came out of the crosscut into the haulage, I found James walking out from the direction of 17C.

"I have not pulled it yet," he said. "But the cleaner sent his boy down with a message to pull because he says it is full to the top."

"Then why?" I could feel the anger lifting my voice. "Why haven't you pulled it. Shit, man, half the shift is over and you haven't even started to pull. You are ok, you just go home in the morning, I have to 'please explain' to the mine captain."

He looked at me and raised his hand. "Just hold on a minute, look around you, John. See these tracks…empty, no loco in sight. There was a derailment at the shaft and nothing has moved here since just after I phoned you. I worked my butt off to get things going again and I have just sent one of the locos through to 16 and I am waiting for the next one, which I will take through to 17."

I sat down at the side of the tunnel and spoke to him. "You knew that you had to pull that box, it was written in the book. If you had sent the loco in first thing it would have had at least one span out by now and the boys would be tipping."

James shook his head. "John, that box is empty. I'm telling you, there is a mistake here somewhere; I pulled that box the night before and I emptied it completely. They have not had time to fill it up again, it cannot be full."

"Bullshit," I said. "The fucking miner's pikanin came and told you it was full just now, you said so yourself."

He shrugged. "We will see," he replied.

We lapsed into silence. My pikanin handed me my cold drink bottle. As I was drinking, I saw James take his own bottle out of his bag. *Well, fuck him anyway*, I thought. When the ore train arrived, I stopped it and told the driver to go to box 6 in 17C. I climbed onto the loco with the driver and James climbed onto the last hopper with the guardboy.

Box 6 was one of the few boxes on the mine that came out of the roof and faced directly down the tunnel rather than sideways to the tracks. To pull it, the ore train moved under it until the first empty hopper was beneath the chute. The guard boy then climbed up onto the platform and opened the radial door by means of a long handle. As the door opened, the ore flowed down the chute until the hopper was full and the gate was closed. Then the loco moved forwards and the next hopper positioned and filled.

The ventilation was good and it was almost cold in the tunnel. The driver stopped the loco just ahead of the box with the first hopper in position and I climbed up onto the platform.

There seemed to be a lot of water dripping from the rock and the normal filthy condition of the crosscut was made worse by the damp feel of the place. James and the guard boy came from the back of the train and climbed up beside me. He pulled a face and shivered but said nothing. The driver of the loco climbed down from his seat and stood underneath us. I remember him lighting a cigarette. I took hold of the handle of the door and tried to pull it down.

For some reason, it would not come and I was turning to the guard boy for help when I was startled by a sudden broad jet of water that came from nowhere and sprayed us all with mud. The last thing I recall before the whole box front seemed to leap out and hit me was the guard jumping to one side and someone yelling.

I tried to turn away but felt a hard blow in the small of my back and then I was driven over and down by an incredible force. There was a searing pain across my face and I was crushed up against something huge and solid. I tried to shout and was immediately choked by mud and water. Then suddenly, I was free and rolling over and over. I hit against steel that I believed was the rail track and tried desperately to grab hold of anything that would stop me.

Movement was so hard, the air seemed thick and heavy and I could see nothing. My hands gripped and slipped, gripped and slipped again. I lost consciousness. When I came to, I was on a stretcher in the back of an ambulance. No one seemed to know anything, and I passed out again to wake up in the hospital.

The next day, the mine captain paid me a visit. He told me that he needed a statement for the enquiry. I closed my eyes and laid back. I did not want to know who had died.

"What happened?" I asked.

He laughed. "You are supposed to tell me that."

I opened my eyes. "Who died, for fuck's sake?" He raised an eyebrow and told me.

"The guardboy. The driver is ok; he had some scratches and a bruise or two. James McArthur is fine, he came out of it with a mouthful of water and a subdued ego; in fact, he did not even go to the dressing station when we got you all to surface. So aside from the guard, you are the one who got the worst of it."

His voice droned on, "Your face will heal, you will have a couple of scars but the doctor says they will fade in time. Of course, your collarbone and ribs will be painful for a while and you will not be able to go back underground for a couple of weeks. You will be on full pay as it was a mine accident. The inspector of mines was underground with me this morning and you are lucky to be alive. The guardboy was knocked into a hopper and smothered."

"You could have gone the same route. McArthur says that you were directly in line with the box door and it must have smashed you through the safety barricade when it burst loose. The mud stretches for some 200 metres down the tunnel…"

Two birds were squabbling on a branch outside the window, and I watched them as he went on about what the manager had said and what the driver had done. Eventually, he took the statement from me and it was as much as I could remember. I signed it and he left. Two days later, my wife brought me a letter from James.

Dear John,

I am sorry to hear about how badly you were hurt. As you probably know, I did not even get scratched. I just remember turning over and over and then I was on my hands and knees and the mud was slowly settling around me. I am writing to tell you that I have resigned from the mine and will probably not see you again. I had a friend once who said to me that the mine was good for him but maybe not for me.

He was right, this is the second time in a year that one of my boys has died and although I did not have to dig him out this time, I have had enough. I don't think that the mine is right for you either. You are one of the few decent people I know in this business but you are also falling for the 'production comes first' syndrome!

I was right you know; that box was empty, and a leaking pipe had filled it up with water. No one said it was full of water, they just kept saying it was full. They assumed that the water was just on the top. I knew it was not full and I should have checked it by going up myself. As usual, there has been a whitewash and no one is to blame. Anyway, it does not matter anymore. I am going to Capetown when I am rich, I will come and see you, battered and scarred by years of rock falls and mud rushes.

Keep well.
James

The Old Man in the Rock

Extract from the law:

A rescue brigade shall consist of not less than five persons in employment at mines, carefully selected on account of their knowledge of underground work, coolness and powers of endurance; No person shall serve as a brigades man unless he has been examined by a registered medical practitioner at least once in every six months and certified medically fit. The manager of the mine may not be a member of the brigade.

Definition of a rockburst:

A seismic event of sudden and violent nature, which may cause damage to underground working and usually results in a delay in production.

After some years outside the industry, during which time I had been a burglar alarm technician, a tracer, a contract draughtsman, a mechanical fitter and finally, supervisor on a multimillion Rand contract, I returned to the mines fatter, more confident and determined this time to make a success of myself. The first night I went underground, I went back to my old job as a nightshift cleaner. I knew nobody but was given to a miner who was to show me the section I was to clean.

I had never forgotten these people who inhabited the mine whilst everyone was asleep and had steeled myself not to show any sign of doubt or sensitivity. I knew that as a new man, I would be watched and any sign of weakness exploited.

After I had changed, I came out of the locker room to which I had been allocated and asked the change house boy for a cup of tea. It had been a long time and it was a different mine but the tea was the same; strong, sweet and stewed but good nonetheless. The cleaners were the same, some looked half civilised, the younger ones tough and aggressive, and most of the older ones were nervous and twitchy. In the corner, his back against the counter was a very large and hard-looking shift boss, probably doing his weekly early shift. I learnt later that his name was Jan Swart.

As I was new, I had a guide for the evening and he was Dirk Esterhuizen; friendly, but worried by something that was connected to me in some way. I caught him watching me a couple of times when he thought I was not paying attention.

Despite some half-hearted jokes and a bit of tomfoolery, we all climbed into the cage without mishap and then with a ringing of bells, dropped into the darkness. I felt at home straight away.

Grunts and shouts of 'Fuck off' soon indicated that this shaft had its fair share of pranksters. There was a loud smack beside me and Dirk flinched, grabbing his ear, someone had flicked it and flicked it hard. This I was to discover was the speciality of Kweekfontein mine, flicking ears and pulling eyelids, both extremely painful and in the ear flicking case often drawing blood.

In a darkened cage, it is almost impossible to see who is responsible. However, if you can correctly guess the culprit, then he will, through some weird sense of honour, stand still whilst you exact retribution. He is allowed to lie and protest as much as he likes but must in the end submit. A sudden sharp searing pain stabbed my right ear, but I did nothing except shift a little sideways and position my feet to turn sharply.

I sensed that the miners in the immediate vicinity all somehow knew what happened and were waiting to see what my reaction would be. The next time it happened, I turned quickly and was in time to see a hand disappearing over the shoulder of the man behind me. I turned on my light and illuminated the shift boss, Jan Swart. It did not look as if I had chosen the best person with which to prove a point.

"Listen," I said, "I am not your fucking playmate; if you want to play your stupid games, then play them with someone else. If you or anyone else does that to me again, I will knock your fucking head off." I switched off my light and turned back to face the front. Someone laughed but I waited in the darkness and could sense no real reaction other than that nobody flicked any ears for the remainder of the trip.

It was an eventful night. The reason for Dirk's concern became a little clearer when we got out of the cage on our level and started the long walk into the working places. There were three of us, myself and Dirk, and someone whom I had not yet met. After an introduction, the other new man, Alan Hengst, asked Dirk which working places was to be my responsibility.

Dirk was immediately on the defensive. "29-53 is my place," he stated flatly and looking at me, "you will be cleaning 29-51a and 32-51."

Here then was the next confrontation. "Well," I said, "I read the intercom book too, and in it the mine overseer says I'm to clean 29-53 and the development." The air became a little thick and Dirk cut it with the flat of his hand.

"My place is 29-53 and I'm cleaning it." Turning, he and the other cleaner walked off down the tunnel. I followed a little way behind.

When we had travelled some considerable distance into the mine, we came to, and passed, a crosscut marked 29-51a. I said nothing and neither did they, we just kept on walking. We passed 29-52, which was barricaded off and then arrived at 29-53. Alan said he would see us later and moved on, but Dirk stopped and, frowning, walked over to me where I stood at the crosscut entrance.

"I see you didn't go into 51," he said.

As he and his friend had spoken Afrikaans all the way from the shaft in the forlorn hope that I, the 'Engelsman', would feel 'out of it', I replied to him in Afrikaans.

"Listen, Dirk, I don't want your places, but I am not going to work where you tell me to either. I am going in here because it was written in the book that I must go in here. If you want to do what you have been told to do and show me the places then fine, if you don't, then you can sit here all night for all I care. Tomorrow, we can see the mine overseer and he can fix it up, but tonight…" I showed him my notebook. "I clean 29-53, panels W1, W2, W3, and W5."

"You can fuck yourself," he said viciously and pushing past me, went back towards the 51 line.

It was not difficult to find my way around; the first panel I went into had two bossboys working there, one of whom, to my relief was the mine overseer's bossboy; a sort of trouble shooter that the mine overseer uses as a second pair of eyes or as a help where sections are having problems with the mining. This big, likeable man was highly respected by the blacks and went by the nickname of 'Madef' because of his thick curly beard. He told me that he spent at least one night a week on nightshift and that he would show me the working places.

The night passed without mishap and I felt comfortable and almost glad to be back underground. I was pleased with the way I had asserted myself and realised that I might have exaggerated the potency of these rough and ready men. If I could maintain the tough image I had begun so well, I was sure that the nightshift would be a pushover. Dirk never went to see the mine overseer as far as I know, and I definitely didn't. 29-53 stayed my responsibility and Dirk was lumbered with 29-51, which I understand consisted of small remnants with very little bonus.

The next night, I met René, my shift boss. The previous night, he had the night off and his stand-in had been too busy covering his own section to visit René's.

René was a breath of fresh air in a dark and dusty hole. He was short, well built and as blond as they come with sparkling blue eyes. I ran into him twelve years later and he had the same bubbling vitality he had had when I first met him. He and I were friends from the first moment.

I had just crawled through a panel and jumped down into the gully when he came down the gully from the other direction.

He stuck out his hand in greeting. "I'm René Dufour," he said.

"Hi," I replied, "I'm James McArthur."

"Let's climb up on the side here," he suggested. "I want to tell you a thing or two."

We climbed up onto the side of the gully and out of the way of the scrapers. He told his pickanin to give him his cold drink and offered me some first.

"Now," he started, "first tell me about the job."

I proceeded to describe what was happening at all the working places, with the exception of the ones below us, which I had not yet visited.

"Do you have a notebook?" He asked, "Did you go into all the places yourself or did you send your pickanin?"

"I do have a notebook," I replied. "But I don't have a pickanin and I always visit every panel myself."

"That's great," he said. "You can call me 'Meneer, Skofbaas'…" He stopped. "Are you English speaking? Yes? Well, you can call me sir, shift boss, or Mr Dufour, but I am not your pal, so you cannot call me René. You must visit each place every night. Tonight, I started at the bottom; tomorrow, I will go in at the top. You must do the opposite and when we meet in the middle, we can exchange notes and then go and visit the places that each of us has yet not seen, at the same time paying attention to the places that the other has warned us about."

He looked at me. "That is my system…" he said, and continued, "…never lie to me, always tell me the truth, even if you make a monumental fuck up and I will protect you. If you lie to me, I will drop you straight in the shit."

He looked around for his pickanin. "Petrus!" He yelled. "Come here, you useless cunt."

Handing Petrus his cold drink bottle, he jumped down into the gully and made to move off. His parting shot to me was, "Don't believe anything these black bastards tell you; check everything yourself, otherwise you might think you are telling me the truth but you won't be." Everything I ever told him about the job whilst I worked for him invariably got the same response, "Did you see it yourself or did they tell you?."

He never had the time to learn to trust me as a miner because only a couple of weeks had gone by when the mine captain, Willem van der Stadt, called me into his office before I left for home late one morning.

"Sit down," he said. "How's the job?"

I sat down and told him I was happy.

"Do you want promotion?"

My heart leapt. *Dayshift*, I thought, *I might be going onto dayshift*.

"I would like that," I replied and smiled. "My wife doesn't like me on nightshift."

"Well, she will have to lump it then, won't she?" He looked at me. "Bill O'Connell in section 22 needs a nightshift shift boss and I told him you could do it. You had better not let me down." I did not know what to say. *Wait until Elsabie hears this*, I thought. She will be over the moon. I knew that the status of the job would more than compensate for the continued nightshift. "Go and report to him now, he needs you to start tonight." He smiled at me as I left, obviously amused by my reaction.

That evening, I walked into the change house after the miners had long gone underground. The first person I saw was René stirring his tea at the counter.

"Hi, René," I said.

"Hey, James, that's great," he shouted and slapped me on the back. He roared with laughter. "Ya, now you can call me René." He shook his head and turned to the shift boss who stood beside him.

"Jan, this is James McArthur, James, this is Jan Swart."

I put my hand out and we shook. Jan said, "I have already met him, he is a cheeky little fucker."

Afterwards underground, I told René what had happened on the first night.

"You are lucky," René said. "Jan used to be an Olympic boxer."

"Even if he wasn't, he would probably tear me to pieces; he is twice my size," I replied.

I later came to know Jan Swart well and as he felt more comfortable with me, he gradually relaxed and resumed his pranks. It was never that he was frightened by anyone but he was concerned. If he felt that he had really upset you or annoyed you, he was always at a loss to know how to rectify matters. He was one of the unfortunate people whose nerves were shot to pieces by the stress of working underground, although he seldom showed it.

Once when someone put a cracker in his pocket, it went off as he was speaking to his mine overseer. Without hesitation, Jan floored the mine overseer with one punch. The mine overseer had to be taken out on a stretcher and Jan was as close to tears as I have ever seen anybody. Fortunately, no one ever told him who put the cracker there, although in fairness, Jan was more concerned about what he had done than by what had been done to him. It was a good team and I enjoyed being part of it.

"René," I asked one day. "How do you get promoted on this mine? How do you become a star?"

René answered without hesitation and very seriously. I saw some of the other shift bosses lean forward slightly to hear him. He held up three fingers and folded them down as he spoke.

"You are in with the management, my boy, if: One, you speak English. Two, you join the Proto team, and three." He paused. "If you fuck black women."

"Of course," he continued, "you will never make it because you are not in the Proto."

"Well," I replied. "As that is obviously the only obstacle for both of us, how do we join?"

We both laughed and Viljoen, one of the eavesdroppers, said, "You Rooinecks think it is funny, this sleeping with blacks, but it is disgusting."

I looked at him and asked if he read the papers, but he got up and left the room. After we had changed and were on our way down, René asked if I was serious about joining the Proto. We discussed it at length and the next morning went to see the Proto Liaison Officer. The first thing he asked about was my glasses. "Can you see without them?"

"Oh yes," I said. "No problem."

"You will be tested," he assured us. "I will put your names on the list; your respective mine overseers must approve and then your section managers. If that is done, then we will arrange a PCT and a medical check-up…including an eye test. If you get through all of that, you are accepted but must then wait for an opening on one of the teams. Then, and only then, will you be sent for training."

"That's that," I said when we came out of his office. "It will take years."

"You had better start worrying about your eyes," René replied. "I know for a fact that there are openings on the teams right now. Let's go and speak to Julian King, he will tell us what the chances are."

What we learnt from Julian was disturbing. Firstly, if my eyes were bad, and they were, I stood no chance at all. Secondly, if you were not fit, you would not get past the PCT.

"What is a PCT?" We asked.

PCT turned out to mean Physical Capacity Test and was calculated by making you step up and down a predetermined size of step in a controlled heat environment. When you had done this for a set period of time, your pulse rate and temperature were measured. Your capacity to endure work under hot conditions was then gauged by some obscure formula that no one could really understand, not even those who applied it.

"One thing is certain," Julian assured us, "if you are not super fit, you won't make it."

A Proto team, otherwise known as a Rescue Brigade, consisted of five to seven members. The work they did was varied and sometimes consisted of rescue missions underground in extremely dangerous conditions. The larger portion of their work was not rescue work as such, but fire-fighting. The South African gold mines have, through the years, developed fire-fighting techniques utilising Proto teams that are the most effective in the world. The name Proto arises from the Greek word meaning first and we were told it was given to the breathing apparatus we used because the wearer was the first person in after a catastrophe.

The members of the rescue brigade were the Elite, known on first name terms to all the managers. They were justifiably big-headed about what they did and I knew that I could give myself no greater boost than to be one of them. After I had got rid of René, I went back to Julian. "Listen," I said, "I need your help. I need to know about this eye test."

"What do you want to know? It is an eye test; like you close one eye and read the chart and close the other eye and read the chart."

"Yes, but then I will fail. Where is it done?"

"The whole medical is done at the mine hospital," Julian replied. "We go every six months. First we lie down and the quack checks us out, then we do the eye test and the PCT and that's it." I left him and went home.

That night, back on shift, I told René, "You and I are going to get fit and I am going to pass the eye test."

So, every morning when we came out of the mine and the other shift bosses laid themselves down in their little nooks and crannies to sleep, René and I trained. They whiled away the early hours waiting for the dayshift by sleeping and we, well we did press-ups, burpees, sit-ups, and ended with a long run.

Those early morning runs in the crisp black of the dying night were fantastic; we talked of dreams and failures, hopes and fears. Aside from Elsabie, my wife, René was one of the few real friends I have ever had.

By the time we were told to report for the medical two months later, René and I were sure nobody could be much fitter. There was only one thing left for me to worry about, the eye test. Fortunately, we were told the day before we were to report where to go the next day, and that same morning that I was told, I called at the hospital.

I went straight to the receptionist. "I have come for my medical for the Proto team," I told her. "Where do I go?"

She pointed down a passage to the right. "Straight down the passage to the end, turn left and it's the third door on the left. Just go straight in and wait. I haven't seen any others, normally there is a whole lot of you."

"Perhaps I am a bit early." I shrugged. "Can I go ahead anyway?"

"Of course, you can." She smiled. "I'm sure they will all be here just now."

As soon as I got inside the room, I looked around. There were some chairs along one wall, a coffee table and some magazines. In the far wall from where I stood was another door. Through that door, I found an examination room and behind the door hanging on a dirty string—paydirt, an eye chart. I stood back and determined which lines I could not read and then copied those lines into my notebook. I quickly returned to the first room and sat down as a man in a white coat entered.

When he saw me, he started. "Can I help you?" He asked.

"I'm waiting for the Proto check-up," I told him.

"Well, you will wait a long time, it's tomorrow."

"I was told today," I insisted.

"No, it's definitely tomorrow."

I shook my head. "It's not my day, is it?" I got up to leave.

"See you tomorrow," he said and smiled.

That night underground, I showed René.

"Ticket to riches," I whispered and stuck the eye test in front of him.

"Where did you get this?" He asked and I told him.

"And if they change it?"

"Oh, fuck off, René. Why should they, but that's not all…" I pulled a small box from my pouch. "Pills," I said. "I figure this PCT thing works on pulse rate, right. Now if we are nervous that won't help at all, so we pop a few of these and presto we are as cool as ice, not a care in the world."

"You can't do it," he said. "You don't know if those things might affect you some other way."

"They are perfectly safe," I told him. "They only calm you down. I've taken them before when I have written exams or gone for an interview."

René shook his head. "No way," he said. "You can if you like, but not me."

"I've got to, René; I must make it tomorrow. I need this Proto thing."

The next morning arrived, and René and I were examined and checked from top to toe. My eyes worked perfectly, reading the letters on the chart from my own internal card, memorised the night before.

The PCT was just scraped through; a score of 31 for me and 32 for René. The requirement was a score of more than 30. Some of the guys were in the 22-23 region and I knew that the pills had not made a difference, the training had.

A week later, we were told to see Leslie Claasens, captain of the 'B' Team.

We were in and we were both on the same team. Training would take place in Welkom and we had to report on the following Monday morning.

Proto training was tough but great. René was nervous and often all thumbs, but for me the equipment, although sometimes very uncomfortable, was like a second home. I had been trained as a scuba diver and moving around in a hostile environment, inside artificial breathing apparatus, was easy.

In the training mock-up, a simulated underground working area, we transported concrete slabs up and down ladders, we performed mock rescue operations in a weird world of foam where the only thing you could see was a billion white bubbles pushed densely against your goggles.

Reality is a strange thing; if an incident is a little unusual, the word that springs to mind is unreal.

A line of six men in single file, tied to each other by safety lines, moving through smoke so thick that the one cannot see the other, even though his hand is on the other's shoulder whilst communicating through a series of hoots from little machines on their chests is unreal.

Doing the same thing down the mine, with the heat searing your lungs at every breath, knowing that if your mask slips, the gas will kill you in seconds, that is real. Believe me.

We were left in no doubt about the hazards. True incidents were related to us by the instructors. Men had lost their lives doing nothing more than building walls to seal off fires. A whole team had taken a wrong turning and become lost, eventually sitting down to die when their two hours of bottled life were over. The importance of this collection of shoulder chafing, nose pinching, leather, rubber and valves called a Proto kit was driven home again and again. We returned to the mine a little apprehensive but proud and somehow stronger.

Proto work is voluntary and although well paid is a sideline. During the course of the year, I went on thirty-one trips for seventeen fires and earned, according to the special tax slip one gets, R6,200.00. The rest of the time, most of the time, I carried on mining.

As planned, my involvement with Proto helped my progress and in a short time, I found myself on dayshift as a shift boss in charge of one of the main transport levels on the shaft, 53 Level.

René stayed on nightshift and we saw less of each other than before.

I had noticed that Bill O'Connell, the mine captain, had been under strain and had associated it with his production shortfalls. I had not had the exposure to tie his production problems to 53 Level before, but once in charge of the level and knee-deep in derailments, falls of ground and jammed ore passes, knew exactly why the strain was showing.

There are various terms for describing the sudden whiplash of violence that ripples rail tracks like spaghetti, hurls tons of rocks hundreds of metres and crushes men and machinery with vicious ease. Rock burst, seismic event and bump are three of them. The experts say it is any failure of rock, which occurs with explosive violence. On surface, it is felt as an earthquake and the mine will tell you that this one was 2.1 on the Richter scale, that one was 3.0. Sometimes they went as high as 4.0 plus.

Underground, if you are far away from the source, it sounds like a pistol shot and is often followed by a tremor or two. If it is close, your world moves, rocks fall, and men stop and wait. They watch the roof, the walls and the floor and then they carry on.

"That was a real son of a bitch," someone might say but the walls and surroundings are still in place so the work continues. Officials underground will move to a telephone to check.

"Hello, Meneer. Where was the bump?"

"Two shaft? It felt fucking close. No, Meneer, everything is ok here. Ya, we will blast everything today. Ok, Meneer, I'll check. Ok bye."

If the bump was very close and reports were streaming in to the surface of closed tunnels and trapped men, all the officials in the neighbourhood would make their way to the area to help in any way possible. The quickest way to get people to the scene of a bump was to send those already underground and close by. Professors scratch their heads and other wise men deliberate and become heated at seminars and conferences. How do you predict a rock burst? It cannot be done they say.

Some areas are more prone than others; some circumstances are indicative of the possibility but we cannot tell you precisely when and where. We wish we could, but sorry…

Down in the pit, meanwhile, we know, but we will not admit to ourselves that it is, like any death, a fact of life. Written down for tomorrow, tonight or next week. In some areas, the mining layouts, the faults and dykes, the conditions all wait but talk to us through one incident after another. They say, "Here, right here, it will happen here, and it will kill, right here." As the months sped by, 53 Level talked to all of us, some listened and left. Some listened but could not leave; others never heard a thing until it was too late.

I worked and worked hard. Aside from the stress of Proto with its long call outs and exhausting physical strain, 53 Level sucked me dry. The main problems arose in an area of not more than 500 metres of tunnel. The area had a double haulage, two crosscuts and a worked-out face above. To further complicate matters, a major fault cut right through the centre of this and was termed by the rock mechanics department as seismically active.

The whole of 53 Level had, however, been neglected and the sixty odd blacks that worked for me found themselves being driven relentlessly to clean, paint, re-lay rail tracks, erect new timber sets, and lash tons of mud from drains stretching over nearly 7 kilometres of tunnel. I must have walked 20 to 30 kilometres each day going from job to job, shouting, swearing, explaining or pleading, whichever seemed best.

The real headache remained. Despite a major effort to remove loose rocks and re-support the seismic area, it moved and pushed as it saw fit. The rail tracks refused to stay level and had to be constantly regraded. The sidewalls pushed inwards ever so slowly but so inexorably that at least twice in my time they reached the point that the ore trains would not pass between them. To solve this, we had to blast the walls out and spend the weekend underground clearing the rubble and reinstalling the support.

All this, only to see the whole process slowly repeat itself. I became irritable and tired and one Monday, nearly destroyed my all-important and carefully nurtured image.

The mine manager was to visit my section. All such visits were planned well in advance so that the responsible people had plenty of time to fix anything that might not be as one would wish it to be.

Bill O'Connell had obviously been boasting a little about his new star shift boss and the progress we had made on 53 Level. It was true that most of the level had been magically transformed. The walls were whitewashed from one end to the other. The tracks were in good repair and levelled as straight as a die from the shaft all the way to the seismic area. Old oil had been put down on the floor to lay the dust and the dull black footwall contrasted impressively with the whitewashed rock of the walls.

Ore trains rocketed back and forth retrieving more tonnage from inside the mine than ever before, but around the corner, deep in the rock, we battled daily with mud and shit. We installed new steel sets in one crosscut, only to find that the last crosscut was now impassable. It was a daily battle over 500 metres of tunnel just to keep the trains on top of the tracks, not to mention filling them with ore from chutes that were pushed out of place and, more often than not, blocked by huge slabs of rock that had peeled off their insides.

I definitely did not need a major visit with its trauma and stern warnings about unacceptable work from someone who last had a decent sweat on sometime before I was born.

I considered laying off sick but instead decided to organise a group of blacks to work on the Sunday afternoon and another on the Sunday night. This would mean fewer blacks on shift on the Monday, which was good. Managers hate to see blacks doing nothing underground. It would also mean a major uninterrupted clean-up of the worst area just before the visit, leaving little time for deterioration.

It worked like a charm. I went underground on Sunday afternoon and worked through one shift with one group, worked through the night shift with the other and then came out to the shaft on the Monday morning to await the arrival of his highness. I had thoughtfully taken a clean overall underground with me and was able to wash my boots and helmet under a tap at the shaft. When the manager climbed out of the cage with his camp followers, I was as nearly resplendent as he was.

These special visits have a protocol like most affairs of state and the mine captain introduced me to everybody in turn, starting with the mine manager.

"James, this is Mr Castell."

"Mr Castell, this is James McArthur, the shift boss responsible for the level."

The manager nodded. "Right, Bill, let's get on with it." He was abrupt and impatient, not a good sign.

Bill was unperturbed and introduced me to the mining manager, Mr Jules Beresford, and the section manager, Mr John Frank. We were by this time surrounded with pikanins, each one belonging to his own particular manager and all armed with cold drink, towels and a change of overalls, in case anybody inadvertently built up a sweat or got wet.

Within a few minutes, we are marching single file down the tunnel led by the mine manager and the show begins. He stops and using his cap lamp which he holds in his hand, he illuminates a rock that is blocking the drain and has obviously fallen off a passing ore train. Everyone stops behind him as he turns and speaks to Mr Beresford. Mr Beresford shakes his head and turns to speak to Mr Frank. Mr Frank turns and speaks to Bill O'Connell and Bill O'Connell turns to me.

"James, this sort of thing causes the drains to overflow which undermines the ballast of the tracks. This is probably why the derailments are above average in this section."

"Right, Mr O'Connell, I'll fix it," I reply.

As I have no pickanin, I pick the rock out of the drain myself and throw it in a dark corner. By this time, the procession, having proceeded without me, is out of sight around the corner. I then hurry after them until they stop for the next bit of mining wisdom. This happened with every minor problem the man could find.

Finally, I catch up to find them all grouped together in my old oil bay.

I am justifiably proud of this place; it is, as far as I know, the only one of its kind. We receive the old oil that we put down on the floor to stop dust, in old oil drums. These are dirty and messy to work with. I had designed a rack down which they could roll with a facility for tapping the oil from them into the smaller drums we use to spread the oil onto the floor. It was neat and clean and had a large sign saying, 'Old Oil Bay'. Everyone else used to stack their drums in some disused tunnel or other.

I waited for some word of praise as Bill told the manager what an improvement the old oil bay was on the previous system.

The manager looked at me quite deliberately and when Bill had finished said, "Your name is McArthur, right?"

I nodded.

"McArthur, what is missing here?"

I looked around to gain time. I had a bad feeling in the pit of my stomach. The silence dragged on to finally be broken by the manager, "Well, McArthur, I am waiting. What is missing here?"

"Nothing," I said finally, my mind a blank.

"You are a shift boss?"

"Yes," I said.

"Well, if you were any kind of a shift boss at all, you would put a no-smoking sign here." He waved at the drums. "This stuff burns, McArthur, and you as a Proto man should know that."

He turned and marched on, his retinue falling into formation behind him, myself at the rear. I was really annoyed and found myself thinking, *Fuck you anyway.*

Further down the haulage, we found six full hoppers in a siding. The procession ground to a halt and the message relay sprang into action. Eventually, I was asked why these hoppers were not being tipped.

On my way to the shaft that morning, I had passed a derailment on a piece of track just ahead. I leapt to the conclusion that the loco for these six hoppers was being used to help re-rail the hopper I had seen earlier. It should have been resolved by this time but it obviously was not.

"There is a derailment just ahead," I said. "I believe that the loco for these hoppers could be assisting there."

We moved on. After five minutes of walking, we reached a long straight stretch of track. The line had been a hive of activity earlier as the blacks struggled to get the derailed hopper back to the tracks. Now it stretched emptily into the distance.

We stopped.

Mr Castell crooked a finger in my direction and I walked over to him.

"McArthur, do you see a derailment?" He asked.

"No, Mr Castell, I don't but…"

He didn't wait for the but, instead, he prodded my chest with a finger, and said, "You are a fucking liar, that's what you are."

I immediately turned and set off back the way we had come.

"McArthur, you come back here. Do you hear me? Come back here."

His voice lacked conviction and I had no intention of going back even if he fired a shot at me.

I went to the shaft, rang for a cage and was on surface by eleven am. I showered and got dressed.

Whilst I was waiting, I wrote a letter of resignation.

Leslie Claasens came to the office during the lunch hour.

"What are you going to do?" He asked.

"I am resigning," I said.

He shook his head. "You won't resign. What about the Proto? You can't leave the Proto."

I did not know what to say, but he was right. All the sweat and tears, all the pushing and conniving, all to be destroyed by some stupid shit with an attitude.

That afternoon, Bill O' Connell took me and my letter to John Frank. Frank read the letter and asked Bill to leave us alone for a few minutes.

"If you were anybody else, you could just go, but you are not. You are somebody that I believe, we all believe, can change things in the mining industry. This game is an old game and very set in its rules and traditions. Mr Castell is one of the old school and believes that it is his job to rule by fear. You saw nothing today, believe me, he didn't even get started." He smiled. "You didn't give him a fair chance."

"Not even my father speaks to me like that," I said.

"Then you are fortunate that your father is civilised, but the mining game isn't. You have got to be tough. You must learn to take it, otherwise, you will never get anywhere. In the mines, there are hundreds of Castells; they shit all over you today and it's forgotten tomorrow."

He leant forward. "Listen and listen carefully. If everybody that didn't like something just ran away, nothing would change. If you think Castell is wrong, then you must work yourself into a position where you can change things and make them right. I have a short temper, I have hit people underground, even miners, but that is changing, more and more people like you are coming into the industry. People with better backgrounds, people that want things to be civilised. Your type can't just leave, you have got to keep coming, otherwise, we will get outdated and die."

He picked up the letter. "I don't have a lot of time; must I tear this thing up or send it through to Personnel?"

"Tear it up," I said. But I knew that even if my resignation was not on paper, it would not be forgotten. The tough James McArthur had taken a step backwards and cracks were showing.

Proto was different to anything I have ever experienced, and it gave me an insight into a world of special men. When such an isolated and select group is allowed to establish its own norms and traditions, it does so with vigour and imagination.

The first thing that I ran up against was finance. We earned large sums of money for services rendered to the mining industry and these monies were paid out to us within a week or two of a fire being brought under control. Whenever the money was ready, our team captain would inform us, we would collect it and proceed to the nearest bar and there sort out who owed what. And owe we did, because for everything that went wrong we were fined.

These fines, when paid, constituted 'the kitty'. The kitty paid for our parties, booze and anything else the team decided to spend it on. Our team had a special meeting within a few weeks of the completion of my training in Welkom.

René and I were to be officially welcomed with a few drinks.

The time was set for four pm, and I strolled into the bar at about five minutes past.

"Hi," I said looking around for a chair, which I found and sat down on. I was to need it.

"You are five minutes late," was the civil reply. "That's R5.00." Leslie Claasens was paging through a small black notebook, finding a new page, he headed it 'J. McA'.

"My wife didn't wake me up," I protested.

"Blaming his wife," someone else chimed in.

"R50.00," Leslie Claasens said.

I was astounded. "R50.00! R50.00 for what?"

"He's arguing," two people joyfully pointed out.

"R100.00," Leslie Claasens said. I opened my mouth and closed it.

"The system that you have just foolishly contributed to is the way we finance our little gatherings," I was told, "and now that you and René have joined us, you are appointed rookies of the 'B' team replacing our friend over there, Nigel Kemp."

Nigel smiled and raised his glass.

"Nigel is one of the best rookies the Proto has ever seen and will do his best to train you to meet the exacting standards he has pioneered."

The man speaking was Gordon Meyer, tall, dark and sophisticated. He was, he often explained to us, here on earth to educate the less privileged, lead the world to salvation and impregnate as many women as he could along the way. He was also reckless in the extreme and the vice-captain of the team. Aside from the later secret telephone calls to my wife, of which he thought I knew nothing, I learnt to admire him tremendously.

"This means of course," he continued, "that Nigel will be fining you constantly, thereby replenishing our sadly depleted coffers."

"Do we get to buy our own drinks from that same kitty?" I asked. "Because if that is the case, I will drink as much as I put in."

Nigel frowned and was about to say something when Gordon interrupted. "A noble sentiment," he explained. "I say that the R100.00 should not be put in the kitty but given to his wife."

"You are a dog," Leslie said, and turning to me. "Watch this bastard like a hawk. This innocent concern for your wife is just the thin end of the wedge. He has caused more divorces than his own and he has had plenty. But yes, it's a good idea, we will give the hundred bucks to your wife. When we next get paid." He wrote in his little book, 'R100.00 for wife…'.

I breathed a sigh of relief. Elsabie would just consider the money ours as all our money was handled that way.

To my consternation, however, three weeks later, after the first trip had been paid for and dues settled, the two of them, Gordon and Leslie, arrived at my house on a Saturday morning. They loaded Elsabie into the car and took her into the middle of town. Here, they gave her the R100.00 and told her, "Spend it, spend it only on yourself, and spend it all."

Of other teams I cannot speak, I never worked with another team, but these five men, these men I knew. Mediocre situations are peopled with mediocre people. Some mediocre people never get a chance to be anything else. I believe that when things become desperate and threaten life itself, then it is not noble men that stand forward, it is the nobility that is dormant in all men that awakens. Everybody has this potential and not one of us has the right to belittle another because in the ordinary trundle of life we can never really know each other.

When deep in the depths of the mine, when the heat is blistering your skin and you are surrounded by thick smoke, when your every breath, despite the sodium phosphate cooler, is burning your throat and lungs and the nose clip cuts the soft flesh of your nostrils, and everything seems to be closing in on you…When the bag on your chest isn't giving you enough air and the bypass won't open because your muscles are screaming at you in panic and fear. When you want to tear the mouthpiece off and gulp great lungfuls of air even though you know it is instant death.

When those things simultaneously surround you and press at you from all sides; the link rope to the man in front and the sound of his hooter, 'advance, stop, retire' are the only contacts with life that you have. How can you not believe in him, this touch of humanity that you can feel at the end of your rope? How can you afford to feel anything but faith in the man who hoots at you, tells you to come or go, stop or start?

No matter what they do in the future or what they become, I will applaud these five men for what they were and what they did then, for me and for themselves.

We were at a Free State Mine, Nigel Kemp, Leslie Claasens, Gordon Meyer, René, myself, and a man from one of the other teams, Anthony Taylor. (Our sixth man was on leave and Anthony was a seventh man on the 'C' team.) I had only been out on three trips before, and this was the first we had been out with Anthony. I was managing well enough but René had shown signs of panic once or twice. When you are new and unused to the equipment, it is nerve shattering to find that your oxygen seems to be cut off or your bypass inoperative.

These things happen, normally because you yourself have made a mistake. To stay cool, isolate and rectify whatever is wrong is the work of a moment.

If your mind won't do it and instead starts to thrash around and shriek at you, life becomes very awkward. The book says, "A man in difficulties may become violent and try to remove his mouthpiece and nose clip. Try to calm him and get him back to the fresh air base as speedily as possible." The crunch comes in the "try to calm him."

The first sign is a long drawn-out hoot and everyone moves to the man in trouble. He points at his set wildly, stabbing his finger towards the offending part. His eyes are large because he was already struggling for some time before he called. You do your best to find out what is wrong but within seconds, he is clawing at his mouthpiece. If he removes it, he is a dead man.

We had our own unwritten system. It said, "One man moves behind the man in trouble, two others move either side of him. If the problem is rectified, fine. If not, the man behind will grip the one in trouble, clamping the mouthpiece hard into position with both hands. The two on either side will each take an arm and keeping the man's hands out of mischief propels him to safety. If safety is too far, the fourth man must then do his best to fix the fault."

We received instructions to enter the smoke and proceed to a holing about 300 or 400 metres inside a worked-out area. There, we were to check the movement of the air and take carbon monoxide and carbon dioxide readings. A fairly routine task and not really frightening. We donned our Proto gear and, standing in a ring, checked our oxygen levels and then showed Leslie our gauges. We emptied the breathing bags and then filled them with oxygen before clipping our noses and sealing ourselves off.

The soft hiss and stop of the non-return valves took over and each in a world of our own we moved off, following Leslie in single file. Leslie, as captain, always went first, followed by Gordon, the vice-captain. I normally followed with René behind me, or the other way around, as long as we, the two rookies, were in the middle. Anthony was behind René, and Nigel, a capable Proto man despite only a year's experience brought up the rear.

It was a difficult trip and used up a good fifteen minutes of our time. We had the same problem we always experienced in worked-out areas; the poor ground conditions and decaying timber support caused by the deterioration over months of disuse made moving awkward.

We entered the crosscut through a ventilation door and about 50 metres from it, turned right up an old travelling way. We could see thick smoke ahead and halfway up the broken and dangerous wooden steps, found matters complicated by a visibility of about 2 metres. Eventually, we reached the top and proceeded up the centre gully and down the strike gully. We were now forced to move at a slow shuffle, visibility had dropped to less than a metre. The heat was becoming intense and we stopped to rest a couple of times. The snot, which always seems to spread everywhere, was burning my face and I bypassed extra oxygen two or three times.

René worried me because I could hear him bypassing air every couple of minutes. When we stopped at the holing, Gordon and Leslie took the readings and I turned to René. I pushed right up close to him and stuck my thumb up in front of his mask; it was a question. His lizard like face swirled eerily in front of me and it nodded. I gripped his upper arm and squeezed. He nodded again. Leslie moved past us and the hooter blurted, "Come on! Come on."

We moved off but I switched link ropes around so that I was behind René. Sure enough, the hissing of his bypass continued. Suddenly, he stopped and started fumbling with his controls. I bumped into him and could just make out his frantic movements. I hit my hooter long and hard and Nigel pushed past me. Dimly, I sensed rather than saw a struggle as they grappled with René and then they were all gone. I was alone. I turned and pulled on my rope, someone was on the end of it.

Anthony appeared from the smoke, he lifted his shoulders; a question.

I got close to him and yelled through my mouthpiece, "Rene." I drew my fingers across my throat to show him that René had no air and then pointed down the gully towards base. He nodded and went past me. I followed on the end of the rope.

Slowly, we made our laborious way back down the gully. When we got to the top of the ladderway, Anthony unhooked the link rope and moved onto the steps, he turned and although the smoke was still thick, I could just see him beckon and then point downwards. Something was wrong and I felt really scared.

The smoke was much thicker than when we had come up and it was very difficult to see where to step or grab hold. We had not been briefed to expect this much smoke. It was necessary to put one foot out, feel for a secure place with my boot and then carefully test it before allowing my weight to move down and forward. Anthony vanished from sight downwards and I kept telling myself that from the bottom of the ladderway it was only 50 metres to fresh air.

Suddenly, there was a loud crash, followed by a rumble of rocks just ahead of me and I knew something had happened to Anthony.

I banged out four hoots and waited. Nothing. I banged them out again, still no reply. I was now really frightened and very, very lonely. Tentatively, I moved forward and found that the steps came to an abrupt end, hanging out in space, what had been a ladder of sorts was no longer. It was hopeless trying to see anything. The smoke seemed thicker than ever; the hiss-stop of my valves was the only familiar comfort around. I sat down on the last step and reaching down with my foot found rocks to stand on. I climbed down and turning around, crawled backwards down the rocks on my hands and knees.

I was shaking visibly. "I don't want to die here," I said out loud.

It seemed that I had climbed for an eternity when I found myself at the bottom of the travelling way. I got to my feet and stood on Anthony.

He's dead, I thought. Even a small fall like that would knock his mouthpiece out and in this smoke, he would be gone in minutes.

I reached down and felt along his body. His bag was still breathing, hiss-stop, hiss-stop. I moved my hands to his face and found his helmet and mask still in place.

As I straightened up, he grabbed my arm and started shouting something through his mask. I bent down to him and saw him pointing down at his leg. I understood only that he could not walk. When I tried to feel down his legs, he kept pushing me away.

I moved off and sat down, my mask hurt like hell. I was short of breath and had already pissed myself. All I wanted to do was run for the ventilation door, but I knew Anthony could and probably would do something stupid if I left him. I crawled back to him and on my knees looked down at him. This was not a joke, this was my responsibility.

I moved behind him and sat down between the rails, bracing my legs against the sleepers of the track. I reached out and gripping his Proto straps at his shoulders, heaved him onto my lap. Then I moved backwards again and heaved. I stopped, panting. I sucked at my mouthpiece and found very little air. I almost panicked before I remembered the bypass and gave myself what I needed. My head was swimming and I felt sick. I waited and when control had returned to some extent, I moved and heaved again.

Sleeper by sleeper, I inched towards the ventilation door. I remember saying over and over again, "Hail Mary, full of grace, the Lord is with thee. Hail Mary, full of grace, the Lord is with thee…"

I have no idea of how long I moved and stopped, moved and stopped, but no matter how long I thought it was, it could not have been more than ten or fifteen minutes. When Leslie and Gordon returned for us, we were about halfway to the door. Within seconds it was all over and I was free.

Out beyond the door, the fresh ventilation was like ice on my open face and I had strong clean air blowing through my lungs. I undid the straps and dropped my gear on the floor.

"You're bloody crazy if you think I am staying in this racket," I stated flatly. "I promised myself in there that if I ever came out alive, I was finished. No money is worth this…nothing is fucking worth it."

Leslie pointed at my Proto set. "If you don't turn off the air," he said, "I'll fine you fifty bucks." I sat down and started vomiting. René turned my set off for me.

Four hours later, after we had showered and were sitting naked in the change room, René came over to me with a brandy and coke. I took it and noticed I was still shaking.

"Are you really finished with Proto?" He asked quietly.

"What about you?" I returned.

He shook his head. "No, I'm ok, I don't want to leave."

"Then I won't either," I said.

The traumas of fire-fighting and rescue work were intermittent. The situation on 53 Level was constant but slowly deteriorating.

Bill O'Connell gave me an assistant, an amusing individual known as 'Stoffel' because of his resemblance to a cartoon character used by the Chamber of Mines in an anti-dust campaign.

Stoffel had resigned himself to 53 Level.

He believed, as only he could put it, "We don't stand a fucking dog's hope, but as long as the mine is prepared to burn its money, I'm prepared to come here for nine hours a day."

He did just that, arriving for work and plodding through the day until he died there some months later.

The first really bad rock burst occurred on Sunday afternoon at three pm. It measured 2.1 on the Richter scale and closed the main tramming haulage for nearly two weeks.

We worked night and day in rotating shifts. We had breaking gangs operating ahead and gangs cleaning up behind them. The clean-up was followed by a meshing and lacing team that drilled and supported the newly exposed rock faces.

After the rock fall was tidied and swept under the carpet, it was possible for a man to walk on top of the service pipes that ran down the tunnel against the roof. Previously, those pipes had been hard against the roof but now the roof had moved up and left the pipes behind.

Despite the concentrated and thorough efforts of all concerned in rectifying the situation, the problems soon returned. The continual battle and the scepticism of my colleagues only served to cement my determination to make things work. I often had two shifts operating when there were tasks that were better performed at times other than in the hectic activity of the dayshift. When the afternoon shift was operating as well, I would sometimes only go down the mine at eleven am and then come out at seven or eight pm in the evening.

This gave me the opportunity of visiting and checking the dayshift as well as supervising the afternoon shift.

What is interesting to me, now on reflection, is that on this level where I had constant stress, I was at times periodically treated to moments of calm and restoration.

From approximately two-thirty pm until the afternoon shift arrived at the working places at about four pm, 53 level was deserted.

From an old disused box chute in one of the crosscuts, warm water cascaded in a constant shower and close by, where a small ventilation leak freshened the air, I established a base. Once the last dayshift man had hurried by on his way to the shaft and surface, I would strip off and shower under the boxfront. Once clean, I would put on fresh overalls and have my lunch.

I often switched off my cap lamp and looked at the dark. There is no blackness like that of the inside of a mine and I found that, if I stared around me intently and for a reasonable period, when I switched on the light again, my surroundings seemed somehow unreal. They sprang into life as if suddenly created from nothing and I would ask myself, was this all here just now in the dark or did it only appear because it had to when the light came on. Every time I was alone down there, 53 talked to me in a special way. It cracked and muttered constantly, whispering, shuffling around as if tired and worn out.

It wanted no more than to be left in peace. Once when Madef and I were struggling to replace a pump in one of the more isolated sumps, the level was talking. He stopped and looked around.

"Listen to the rock, Boss James, it is talking. The rock is calling us, it wants you and me."

I too stood and listened. "For you, it says come," I replied, "but to me it says, 'Leave me alone'."

He shook his head. "That can never be, the mine wants gold, as long as there is only a little bit of gold, the mine will come after it, just like a dog that wants sex."

Those lonely hours were never truly lonely; there was always something there that seemed real and alive, something that had no need of hate or envy with which to find its vitality. Something that nonetheless was no friend, rather like a machine, it could serve or destroy, you had only to understand the rules.

We understood each other, that presence and I. We were both quite content to rest and contemplate trivia when no one else was around. Once the crews arrived, it was over; wearily I would get to my feet and start to push, 53 Level would start to push back and the struggle would resume.

I stayed, the majority of blacks stayed, Stoffel stayed, others came and went. One young learner miner in the last months of his training decided that mining was not worth the money. After three shifts, and quite suddenly, in the middle of the morning, whilst we were working on some supports, he made his decision and told me, "This place is a death trap and you, you of all people, should stop it."

"This is my job," I replied, "but if you can't handle it, it's not yours. Either you get back up here with me or you fuck off out of the mine."

He laughed. "You think I am staying here in this?" He turned and walked off.

As he went past Stoffel, he pointed over his shoulder at me and said to him, "You go and help him. You're as mad as he is."

He went to surface and resigned. We never saw him again. We got on with the job. The constant strain took its toll however, and personalities changed. There was less time for the good things and baser instincts came more easily to the fore.

I have been hard underground; I have been inconsiderate and had a fair share of complaints and grievances directed against me, but I was never violent and very seldom, if ever, attacked anyone's personality or make-up. This too, soon went by the board and I found myself charged with assault.

I was in search of a pipe wrench and was sure that I had seen the pipe fitter's crew turn into 53/37 crosscut that morning. I left the boys busy with preparation work and set off to see if I could borrow the wrench from wherever it was that they were working. I went myself because I knew that they would not lend it to another black but that my position would perhaps swing the scales in my favour.

I was halfway down the crosscut when I saw them busy fitting a manifold onto the end of one of the columns. As I passed beneath a boxfront platform, I heard something shift above me.

One of the boys was sleeping on the platform as if there was no mine, no problems and only his comfort was of any importance. I reached up and grabbed his leg and pulled, he fell to the floor beside me with a crash and no sooner hit the ground, than he was on his feet and running. He must have thought himself in the middle of a nightmare because he screamed and flung his arms up, his eyes rolling wildly as he took off down the haulage.

"Come here!" I yelled at him, and he turned, picking up a rock as he did so. Before I realised what his intentions were, he threw it, striking me painfully on the upper leg. He turned to run, still shrieking wildly but slipped and fell. I watched as he got to his hands and knees and then groped for another rock. This time, I was not giving him a chance. Who the hell was he, goofing off and then throwing rocks at me?

An unreasoning anger filled me and I rushed at him. When I reached him, he was just about to straighten up. As he got to his feet, the rock still in his hand, I hit him. I know now that he was frantic with fear and had no idea at all of what was happening, but at the moment I hit him, to me, he was vicious and deadly…he was the focus of all the stress, all the failure and futility of my work. I hit him so hard that my arm went numb from my fist right up into my shoulder. The pipe crew grabbed me from behind and the boy fled sobbing hysterically down the tunnel.

Later that afternoon, a shift boss came into the office.

"Where is McArthur?"

I emerged from Bill O'Connell's office and presented myself.

"Did you fuck up my boy on 53 level this morning?" He asked.

"Well, I would hardly put it that way," I replied, "but yes, I hit a boy this morning."

One of the other shift bosses rubbed his hands together. "At last, McArthur hits boy, leaves mine in disgrace."

I ignored him; this was serious and could have far reaching consequences. The shift boss, Duvenhage, confirmed my fears.

"Hit him, oh no, you didn't just hit him, McArthur, you put him in hospital. You broke his cheekbone and dislocated his jaw."

I stood there disbelieving. "He was ok," I said. "He just ran off down the haulage. I only hit him once."

Then I remembered the fall from the platform. "He fell," I started to say. "He fell from the platform."

"Bullshit, McArthur! You fucked him up with a pole or something and I am personally going to lay an assault charge."

He stormed out of the office.

Bill O'Connell called me back into his office and asked me what had happened. I told him everything.

"Leave it with me," he said. "I'll see what I can do." He looked and sounded tired.

"Do you want some time off?" He asked.

"Impossible," I replied. "I've got too many things on the go; maybe in a week or two but not now."

The next day, I was called to see the senior personnel officer. "I have here a very serious allegation. Do you know what it is?"

"Yes," I said.

He waited until he saw I would add nothing and then continued. "This kind of thing carries a possibility of instant dismissal and quite frankly, the medical report supports an attack of particular viciousness."

I had spent a fairly sleepless night torn between remorse one minute and a feeling that the mine must do its worst the next. I was tired and not very pleased with myself.

"Look," I said, "what happened, happened. I am not going to deny it and if you want to hold a full-scale inquiry, then go for it. The bastard is not exactly innocent either. He was throwing rocks at me for fuck's sake."

"Yes," he said. "So it appears, I sent someone underground this morning and there are witnesses." He seemed reluctant to come to the point and I waited.

"The management feels that I should not pursue this too diligently, but I must weigh all the facts; the man's shift boss wants you charged."

"I don't want to be charged," I replied, "but in all fairness, I did hit the guy and if I have to pay…" I shrugged. "Why don't you ask him if he wants to lay a charge?"

"Are you prepared to make a statement admitting that you hit him?" He asked.

"I'm prepared to make a statement saying what happened," I countered.

He proceeded to take a statement in which I detailed what had occurred as honestly as I possibly could.

Two days later, Bill O'Connell came in the office and said, "That boy you hit is not laying charges."

"And Duvenhage?" I asked.

He shrugged. "Duvenhage wasn't the one who got hit, he'll cool off. Van der Stadt won't countersign any paperwork charging you with assault. One shift boss charging another for assaulting a black is not the way we want to operate on this mine."

That was the end of that. Duvenhage got over his indignation and I never saw the black again.

A few weeks afterwards, at about six-thirty pm one evening, Bill O'Connell, myself and a shift boss called Nicholas Roderick were sitting in Bill's office, drinking. This is something that occurred very frequently, much to my wife's disgust. It had become so bad that several of the wives had got together and complained to the management asking them to forbid alcohol on company premises. Everybody denied that their particular wife would ever do such a thing and maintained vigorously that they drank as much as they liked, no matter what their wives said.

However, for a while, the band of merry men was sadly depleted. The die-hards (mostly our section) stuck to their guns, or rather their bottles, and the no alcohol regulation became one of the many that no one took any notice of.

Bill had his feet on his desk and I had just got up to replenish my glass when the whole building shook violently. We waited for the aftershock that would indicate that the bump was elsewhere, but it didn't come. Bill was on his feet instantly.

"That's 53 Level," he exclaimed. "James, who is underground?"

"I've got two loco crews pulling boxes," I replied as I went through the door.

With very few people underground, if 53 had bumped, they would probably be isolated and in a great deal of trouble. Neither Nicholas nor myself questioned Bill's judgment and the three of us were changed and on the bank in minutes. The cages, however, take time and it was not until after seven that we were at the 53 Level diesel bay. Two of the locos were missing and we hurried inside. I have never been able to walk as fast as the others and as usual, I was at the rear, heart pounding with fright and head swimming with booze.

We had not gone very far when we saw a light approaching, flickering up and down. The helmet it was on obviously belonged to someone running and leaping along the track. We stopped and waited. Nicholas was the first to realise that the worker running towards us had no intention of stopping. He leapt at him and tried to grab something tangible and succeeded in grasping his overalls. Bill and myself watched the fight that ensued in amazement. The unfortunate black jumped and pushed, the two of them whirling in circles down the tunnel.

Eventually, minus helmet, with his cap lamp bouncing along the ground behind him and his overall torn to the waist, our wild man broke loose and resumed his headlong flight.

"Stop him," shouted Nicholas unnecessarily, for the escapee, stumbling on his lamp, measured his length on the floor. Nicholas and I were on him in an instant. We sat on him until his energy was spent and then slowly but surely calmed him down.

I recognised him as 'Basjan', one of loco drivers. Eventually, we got some sense out of him and found out that he had got off his loco to drink water when the world caved in. His loco had disappeared under tons of rock. He had no idea of what had happened to his boy or to the other teams. We had no alternative but to send Nicholas back with him whilst Bill and I went on alone.

The damage was immense. Before we even reached the area where the rock had closed in on itself leaving little or no tunnel at all, we found whole lengths of track lifted and tossed sideways against the sidewall. In one place, a crater a metre deep and 3 metres across was surrounded by shattered rock, no piece bigger than a dinner plate. A pile of cement bricks that I had offloaded and packed into a small cubby only two days previously was spread over a length of at least 30 metres.

Bill and I tried to penetrate as far as the loco in order to check on the guard boy and the other team member, but it was hopeless.

"How much further in could Basjan have been?" Bill asked.

"The nearest drinking water is still about 60 metres in," I told him.

"How in the hell did he come out through this?" Bill shook his head. We looked at the time, it was just after eight.

"We will have to go back and get some help," said Bill.

Reluctantly, we set off, back down the tunnel, headed for the nearest phone. Before we had gone far, we found Nicholas approaching with Madef and an assortment of blacks. He told us that the fall could only extend 100 metres or so as the guard boy and the other loco team were uninjured and had found themselves on the far side of the bump when it had happened. They had been sure Basjan had been killed as he had gone to drink water and then the rock had come down to separate them, for what they thought was forever.

They had, in fear, retreated further into the mine and then climbed up through one of the working places to the level above. From there, they made their way to the shaft.

No one had died; no one was even seriously injured. How Basjan had fought his way through all the jumble of broken rock and ruptured supports none of us knew, but he had done it in one piece and, after a day off, was back at work as if nothing had happened.

What had happened was that the tired and impatient rock had spoken to us for the last time.

Elsabie, my wife, was ever present, ever supportive, but as time progressed and I became irritable and inconsiderate, I began to sense that she too was feeling the strain. I would lose my temper over minor inconveniences and storm out of the house in self-indignant rage. The extra money that I earned through my Proto activities never seemed to do anything concrete towards stabilising our finances. The amount of time I was away and the many drinking sprees I indulged in with the team did plenty to undermine our relationship.

I was so much on edge towards the end that the sound of the telephone would visibly shock me and Elsabie was worried. On several occasions, if she felt that I was receptive, she would suggest that I slow down.

I went so far as to ask Bill O' Connell if I could transfer back to nightshift. René had never left it and seemed to bounce along very nicely. I remembered the good times we had with nostalgia and could see myself sitting back and running a cleaning contract with ease. The proposal was laughed out of court; I had no chance of being allowed to leave 53 level. Then I had another close call on the Proto and realised that I must drastically reduce the pace of my life.

It was one of those long boring trips on which nothing happened and all we were required to do was check on an already sealed fire by taking periodic readings in the chimney. A chimney was the gas and fume outlet for an area which had been on fire and had been sealed off to force the fire to smother itself. The gas readings in the chimney indicated how intensely it was still burning. This fire had been successfully sealed and Nigel and I were assigned to take the readings two days later. These trips did not even require full teams and it was normal to be called out in pairs.

Nigel was asleep and it was my turn to do the readings. I got up and pushed my way through the vent door. I was tired and I propped the door open to save myself some effort later. I took the reading and made the necessary notes in my book. I began to feel a little uncomfortable and sat down on a pile of wood for a few minutes. I felt giddy and leant back against the rock, then realised in shock that I was showing symptoms of gas poisoning. I hastily got up and made my way to the door. My head started pounding and the pain became intense. I woke Nigel and told him what had happened.

Half asleep, he looked at me and asked, "Where is your set? Why don't you breathe some oxygen, that should help."

He stopped and stared at me incredulously, then was suddenly fully awake, "You fucking idiot!" He yelled, getting to his feet. "You didn't have your set, you didn't wear your fucking set."

We got my Proto set and I pumped pure oxygen into my lungs for a few minutes. After a while, I began to feel better or at least I told Nigel that I did, I had no intention of letting him report this to 'control'.

I had a headache of similar proportions once or twice before from explosive fumes. This headache was not the same. I had bought it on myself through inexplicable stupidity; I had once again made a serious mistake and badly frightened myself. I could so easily have sat on that pile of wood and died.

When I received a phone call from Cape Town a week later, it was a message from heaven. I had, through the auspices of a concerned person, been given a chance of a job in a factory in Bellville. I contacted them immediately and arranged for an interview the following Friday. I was very sure from the circumstances of the offer and the reaction of the manager over the telephone that the job was mine. I went into work the next day and gave twenty-four hours' notice.

I was flattered and surprised at the reaction. Once people realised that I had no intention of staying, that I really was leaving, I was greeted with pleas and advice on my foolishness everywhere.

Two of the appeals made sense; Leslie Claasens called me to his office and tried hard to talk me out of it. Bill O'Connell did the same.

René, however, was pleased. "You lucky bastard," he told me. "You must organise a job for me too."

The next day, I went in to work for the last time and we had the usual celebration.

"Any excuse for a dop," was Gordon's comment. He took me aside afterwards and told me confidentially that it was a wise move as Elsabie was on the verge of going to bed with him.

"It would destroy you," he told me knowingly. "At the moment, she believes you to be the best, but if…"

He wandered off leaving me supposedly very relieved at my sensible decision.

And so I left. It was that simple. What more can I say?

In Cape Town, three days later, as I was being told by an apologetic manager that I had overreacted to his offer and that the job had been given to someone else, my old man in the rock lashed out and took the lives of over thirty-five people in the space of seconds.

Numbered among the dead were the shift boss who replaced me; Jan Swart, the pipe fitter who worked that section of the mine, Stoffel, Bill O'Connell and another mine captain, Willem van der Stadt.

They and a team of thirty or more blacks had gathered in just that place where the rock and I spent our lunchtime truces. I understand that they were to make a major attempt to solve some of the problems that had beset me during my long struggle.

The details of the accident and the subsequent heroism as well as the total commitment of every fibre of the mine to a rescue that lasted days after all reasonable hope was gone are recorded elsewhere. The mine closes its ranks to outsiders at certain times and in this case, I too fall silent. The story is too personal and involves the loved ones of too many already injured lives to tear open and expose.

I hitchhiked back to attend the memorial service. René was the only one to receive me as someone who still belonged. He and I sat silent for a long time after he had told me all he knew. His involvement had aged him and he was bitter in his condemnation of management and God alike.

I knew that haulage. I knew its physical and human failures. I also knew the man in the rock; we had fought each other and fenced to a private set of rules. In the end, I ran away and he, well perhaps, he showed his contempt. Deep down in the mine, he reached out and marked me thousands of miles away. That single touch changed my attitude to life forever and I have felt ever since that I should have been there, not to die, I do not want to die, but to share.

To somehow say to my crew and friends, I am here too, if it makes any difference. I am here with you.

Part Two

The Creation

Someone recently said this to me, "Such a place as you have described could easily be the origin of a legend similar to that of King Solomon's mine."

There are no facts that support this. There are, however, emotions; emotions of awe and humility that men have felt when confronted with a creation of divine proportions. Emotions of greed and selfishness when struck by the magnificence of riches far beyond the normal allocations of their lifestyles. I, who have experienced all these, can say with conviction, "Yes, this could have inspired a legend, this could have brought men from far and wide, full of hope and destruction."

About the time that the Roman civilisation was feeling the fatal thrust of Barbarian swords and Gildas was writing his *Ruin and Conquest of Britain,* a less threatened and far simpler people were mining copper from a 'Green Mountain'. They used the metal for trade and, as is normal in such matters, continued doing so quite happily until the Europeans arrived and took over.

However, production requires more than just a scratching of the surface. The pursuit of riches ensured that the mine expanded into a town and deepened into a network of shafts and tunnels. Instead of a dusty lean figure patiently digging at the ground in the shade of a wild thorn, we find a sprawling town and torn nerves in the depths of the earth.

But this mine had a longer history than that. In fact, men were still standing in the wings and would not come on stage for six hundred million years when the first twinges of labour started to disturb the womb that was to carry this unique creation. A long fracture extended itself from the earth's surface to many thousand metres below and waited. Water from surface, finding an underground passage, slowly filtered its way through this new avenue and in its timeless exploration took away with one hand and gave with the other.

The water invaded with erosion, both chemical and physical, it also colonised by deposition, both chemical and physical. It carved great cavities and small fissures, filling them with sediments from other distant sources. The patience of the earth was not as endless as the water however, and it intervened, sending wild heavings of heated rock and fluids up from below. These forays of hot and relentless plunderers changed the nature of all the inhabitants in that slow, aged community. New minerals were born, and others were transformed. The water dripped and filtered its way through fresh materials, strange pockets and offshoots that had not existed before.

Then the earth, bored with its new toy, receded but the water stayed and when all was quiet and calm and the water could hear only itself, it pondered a new creation. *Look around*, it thought, *look at the lead and the zinc, at the copper and germanium; here are the riches of the world, here is peace and time, endless time. Here will I create flowers from stone, rainbows from sand and grit; here will I build palaces of glass and light, the like of which have never been seen or ever will be seen on the surface of all the world.*

In later years, two men confronted one another in an office directly above this patient masterpiece. One was, for want of a better term, its owner, the other its agent. Neither of them had more than fifty years of existence to their credit and neither of them could nor would create anything that would see out the end of the century. They both had other abilities; the owner the ability to formulate a policy that would destroy every flower and rainbow that existed below his feet and the other the ability to sell and distribute those same works of time to the world. How much either of them felt for the charge placed in their hands is hard to say.

Through lineage, Solomon of biblical fame was represented in one, and Eric, a Viking, in the other.

Solomon had been ushered into the office by a secretary. Once an engineer on the mine, he knew the other well. Eric did not get up but suggested that Solomon sit down.

"If you want your job back, it's out of the question," he said.

"No, I don't want my job back," Solomon replied. "I want to talk to you about crystals."

Eric was short. "I have no wish to discuss crystals. In fact, if that is all you have to see me about, we can terminate this conversation right now." His anger was thinly disguised by his business-like tone.

Solomon was not perturbed and continued calmly, "I would very much appreciate a few minutes of your time. I would like to explain to you a facet of this business that you and the company you represent are missing."

Eric leant forward and deliberately interrupted. "The only facets of this business that interest me are the loss in production that occurs whenever someone finds a pocket, and half the bloody mine stops work to go and dig there. The costs of having some smart ass drill and blast metres of rock totally off direction because he believes that he knows where the find of the decade must be situated. The production delays caused by miners holding up their blasts so that the sequence of mining is more favourable to them than the other shift."

"I have a job to do and I operate as best I can to ensure that the shareholders of this mine are glad that it is me sitting in this office. The profit that you and people like you are making by exploiting the illegal removal of minerals from the mine is one of the things that I am expected to stop." He paused. "And stop it I will, believe me, Mr Solomon."

Solomon raised his hand. "Mr Eric, I am aware of your problems and I sympathise, but I must point out to you that the mine, and therefore, you, have a civic and a global responsibility that is as important as your financial one. You are destroying an aspect of our heritage that can never be replaced or recreated."

Solomon pointed to the display cabinet where an Azurite crystal cluster formed the centrepiece of an impressive display. "That piece is worth at least R2000.00 at any rock convention in the world, but it is worth many times that as a creation of God. It is unique and can never be reproduced. If I placed a cheque for R2000.00 on your desk and smashed the rock, would that be just or right?"

"The money is not the issue, we buy and sell using money, it is the revenue we earn for recovering and distributing the article and that's all. Only a fee, a percentage. The value of what we handle, Mr Eric, cannot be assessed in financial terms."

Eric rose to his feet and walked to the cabinet; he looked at the crystal for a long time.

"The other day," he said, "we reported a Mr van Biljon to the police. He is a miner on 32 level. When we opened his garage, he had shelves of stones along every wall, and a glass topped table in the centre of the garage for people to examine his crystals. He had special lighting to best show off their worth. I understand his income from buying and selling stones is phenomenal. He beats his wife, drinks himself into a stupor whenever he can and has a police record. Do you think he gives a damn about God's creation or global responsibility?"

He returned to his desk and sat down. "Stop wasting my time, Mr Solomon. You are fortunate in that you are a licensed dealer and as such, we both know, almost impossible to prosecute. I have to accept you as part of the environment. These men that work for me, and work for me they must if they want access to crystals, they are a different matter. I will probably never stop them completely, but as far as I can, I will protect what I feel are the interests of this mine in the ways which seem expedient to me."

Solomon got up and thanked Eric for his time. "You win." He smiled. "But I have a small gift for you." He passed a thin book across the desk.

"What is this?" Eric asked, surprised.

"It is a recent publication I received from one of my American buyers. It was on sale at the Annual World Mineral Convention in Texas. It is called *Green Mountain: The Unique Ore Body*; it is about this mine, my friend."

Eric opened it and seeing the pages of colour photographs of crystals, closed it. He tossed it back across the desk.

"No," said Solomon. "Keep it. I have arranged for a further fifty copies to be delivered to the head office of this mine in New York. I feel that we should all know what it is that we are dealing with, don't you?"

The mine has never officially changed its policy and even today, there are random searches at the gates. The disciplinary code states that if you are caught removing crystals from the premises, you can face instant dismissal. To my knowledge, this has never been the case for a first or even second offence. Crystals are a part of life in the world of Green Mountain. Private collections the world over include many beautiful examples of stones carefully removed and packed in dark and fume-laden holes underground.

These same stones that grew silently under the patient care of the creator for hundreds of thousands of years. However, for every one that is safely removed, there are thousands that are crushed, broken or swept away by mechanical loaders and conveyor belts.

How strange it is that people that will spend millions to erect a concrete and glass replica hundreds of floors high, will not officially create a system to preserve the finely worked and perfect original. I saw many things in the mines to either break down or rebuild my faith in men. In the depths of the 'Green Mountain', I found faith in the magnificence of creation.

A Pocket Full of Stardust

The shift boss' office on 26 Level was a place where fortunes were made and spent. It was a small dirty area of the mine, which had spawned greater dreams than the coffee bars of a stock exchange. Poor men, battling daily with where the next bottle was coming from, heard and told tales of incredible finds, of pockets of crystals so beautiful and rare that the mines and debt collectors became forgotten, never to be seen again.

I was unimpressed.

Yes, I knew about crystals. I found small pockets of beautiful stones regularly in my working places.

Pink Smitsonite; a rare crystal listed as being found as white, sometimes blue or green. In my working place, a delicate pink rose of incredible value. I had sold at least a thousand rands worth over the last six months. But retire? No crystal pocket could produce enough to supply my wants for even a year, let alone the rest of my life. So I listened, but with half an ear, to the tale of a man who two years previously had found and removed enough Azurite to buy a thriving business in Cape Town. About the rich landowner, once miner, who welcomed visits from his less fortunate fellows who were still working in the bowels of the earth.

Then there were the other fortune seekers that had come from oversees in search of similar specimens to those they had seen in collections of richer men. Young and old, their stories were repeated endlessly.

"Remember Barry? Fuck, what a jerk. He wanted to get a collection. He used to sit right there where you're sitting. No shit, just there. The boys fucked him around, everybody fucked him around. We used to call him 'Cockroach'. Do you remember 'Cockroach', Dougie?"

"Yeah, I remember him. What about that time he ate that sandwich? Who gave him the sandwich?"

"It was Tyler. Tyler came in here and put the cockroaches in Claude's sandwich and Claude gave the sandwich to Cockroach. Shit, we laughed; the fucking little cockroaches were trying to get out after he took a bite and one of their legs caught in his moustache. Shit, what a scene! There were tables flying around and coffee all over the place. John's books were all on the floor."

Dougie laughs and slaps the table but John interrupts.

"Well, he said he was only going to stay until his rock collection was complete. You have to admit, he was a clever bastard; he could work anything out; if you wanted to know something, like how much interest you must pay on your car, or what the tonnage of your blast would be. No problem, he used to just work it out."

John turns to me. I am the self-professed sceptic, the one everyone always had to convince.

"He didn't come to the office one day and we all thought he was having problems, you know big rocks to blast or something. Afterwards, we found out from the boys that he found a pocket of Dioptaz with big crystals growing out of a calcite bed. The sneaky little bastard carried it all to the sub-level above, which was worked out, and hid it. The boys told us it took him a week to get it out of the mine. Then he resigned and we never saw him again. He probably lives in a house on that special street, you know where all the prossies sit in the windows and wave at you with their tits."

"How much do you think he got?" I'm forced to ask.

Dougie tells me, "Old Solomon reckons he paid out R25,000 and he only bought a few pieces."

John shakes his head. "His bossboy paid cash for a fucking great station wagon, so just work it out."

It was true that many of the employees on the mine were not run of the mill mining men. Many were just as described, youngsters from mining colleges or gemstone collectors. Wildly dedicated to the earth and its treasures, the wonder of the creation at Green Mountain and its showcase of precious pieces drew them like flies. They came from everywhere, sacrificing a period of their lives for the chance that they might possess some of nature's deeper favours.

I had certainly not come in search of riches; driven by need and fear, I had sought what I believed to be the safest mine I could find. An old friend of mine had told me, "Green Mountain has not had a fatal accident in twenty-five years." He was wrong but close enough.

I found myself working as a miner with a small section and little responsibility. The work was tough but not stressful, the pay was fair, and I was happy. I was, however, human and as was the unwritten rule, entitled to whatever crystals occurred in my working place. I soon became adept at recovering them from the little hollows of rock in which they had grown over the ages.

The normal course of events on a mining shift started with examining the working area, you then went on to make it safe and then got your crew to start work. The first thing they did was clean the ore from the previous blast in preparation for your own blast. Whilst this was in progress, I would go over the rock face with a chisel and a hammer, probing fissures and cracks, sounding any area I believed might hide a hollow or pocket. If I was fortunate enough to find a reasonable pocket, it might mean anything from half an hour to four or five hours of solid work to chisel out seven or eight good specimens.

For a specimen to have value, it had to be firmly embedded in the rock from which it had grown. To extract it meant chiselling the rock it was seated on. Sometimes the crystal was so sensitive and fragile, that too sharp a blow on the hammer, even a foot away, would shatter it. Once released from the rock, you then had to pack it in such a manner that you could drop it, fling it over your shoulder or allow it to be kicked by a stumbling miner with no fear of damage.

If the police boys saw you being over protective of your bag, they could and probably would search it. Even if the other miners realised that you had something, it created possibilities of begging or even theft.

I had two identical bags one packed into the other. Rolled into the second was my kit; a hammer and chisel, a roll of toilet paper, a bag of sugar and several plastic containers, which fitted one into the other. If my finds were reasonably sturdy, I would wrap them tightly in toilet paper, winding it round and round the stone until it was a tight ball of tissue with the rock in the middle. The very fragile stones I would pack into the plastic containers. Into the bottom of the container I would pour a couple of centimetres of sugar, on top of the sugar I would carefully place my crystal right in the centre of the container.

Then I would fill the container with more sugar until the crystal was covered and the container full. When sealed, the container would protect its tightly packed contents perfectly. At home, I could dissolve the sugar under a tap. In this way, I recovered and sold hundreds of exquisite specimens, which probably grace collections all over the world.

However, I only earned a really large sum of money from crystals once. But that once was enough to show me that the stories I had been hearing were not as farfetched as I had believed.

As I write, I have before me a list of minerals that are found in the 'Green Mountain' deposit; it is two pages long. With such a wealth of material at hand, nature was able to group colours and textures in a variety of ways. The more identifiable crystal types found on a single base, the rarer and the more valuable the piece. In this particular instance, no less than four different crystals were identifiable, from a sphalerite base to a final sugar coating of fine yellow mimetite crystals.

I walked into the underground office one evening to hand in my production figures. I found only Basil there poring over the intercom book.

"Where is John?" I asked.

He didn't answer but only shook his head.

I sat down and started on my lunch, opening the bread and checking carefully on the contents before eating.

After a period of silence, he looked up and I asked him, "Where is Claude?"

"I have not seen anybody except you," he replied and took one of my sandwiches. He pushed his glasses up onto the bridge of his nose from where they continually slipped. "Come to think of it, John didn't check my place today."

"He didn't check mine either," I said.

We smiled at each other. "Perhaps they are both so busy in Claude's place that they haven't had time." As he spoke, we got to our feet.

"Well," I added, "I think a little walk might be in order. I actually want to ask John if he will order some primers for me."

Basil thought he might have an important question for John too, so we set off at once for Claude's working place, both pretty certain as to what we would find there. Before we turned the corner into the drive leading to Claude's place, Basil stopped me and turned off his cap lamp. He motioned me to stay where I was and walked the few steps to the corner in the light from my lamp. He leant forward and looked into the drive. After a few seconds, he motioned me closer.

I crept up to him and looked. Halfway down the drive where the ladder to Claude's working place disappeared into the roof, Claude and John were busy setting out specimens all over the floor.

There must have been at least thirty or forty of them, some small and some as big as dinner plates.

Basil and I both stepped out into the drive absolutely amazed.

Basil was pushing his glasses up his nose so often, his hand never left his face.

"Oh man, look at that," he breathed.

John and Claude both looked up and saw us at the same time. Claude had stripped his overall to the waist and they were both sweating profusely.

"Now that you know," John said, "you might as well help."

He paused and looked at Claude and then at us, "On one condition, you accept what we give you and that's all you get."

"Fuck you," Basil replied. "We do it the proper way. After we have got everything out, we put all the stones around like this and then Claude chooses first because it's his place, then you because you're the shift boss, then James and I choose. After that, Claude starts again and so on."

John shook his head. "Go with Claude and let him show you, then if you want in, you take what we give you."

Claude turned and went up the ladder, and we followed close behind him.

Once in the working place, we stopped. All around us were specimens; big, small, broken and whole. Above us, on top of some of the supports, sat a black holding a large rock covered in crystals. He handed it down to Claude and then got to his feet, the top half of his body disappearing into the roof.

"Hey, Petrus," Claude called, "climb down, I want to go in."

To our astonishment, not only did Petrus climb down, but another one of Claude's boys emerged from the hole and dropped to the ground.

Claude laughed out loud and bowing, said, "After you!"

I went up like a monkey and emerged in a pocket that was adorned with such beauty and splendour as to rival anything I had ever heard of.

The pocket was the size of a small room, but much higher. It extended at least four metres up into the rock. The whole of it was an undulating mass of silver white crust-like plates covered in pale green frost. It looked like a Swiss landscape on a box of chocolates. On top of this were deposited lead crystals, fine as cut glass. The clarity of the crystals and their perfection sent the beam of my light rocketing back and forth around me. Everywhere I looked, the beam of my cap lamp followed and where the beam touched, it burst into a thousand slivers of light.

It was only after a few minutes had passed that I realised the yellow in this rainbow of colours was real and that many of the lead crystals were topped with a coating of mimetite, no single crystal larger than a grain of sugar. It was magnificent and I wished then, as I wish now, that such a place could be preserved and transported to a great centre somehow, so that the world could see what artistry lies hidden, deep from the sight of men.

For whom are these things created?

Why is it that they lie hidden beneath tons of rock? And why, the greatest why of all, why are we, specks of dust on the surface of an ocean of time, allowed to blast such a masterpiece into flying debris?

Humbled, I climbed down and Basil went up. John appeared from the back of the working place and when Basil climbed out of the pocket, we sat down under the hole and listened to John as he outlined his plan of action.

"Firstly, you will have nothing to do with the digging out of the stones. The mining must go on in the section; nobody must even think that there is anything happening on this shift that is out of the ordinary. Secondly, the best stones must be dug out tonight and hidden so that we can make proper arrangements to get them out of the mine. Thirdly, Claude and I will be the only people who will sell any one of these stones. There will be no other source except us, and the dealers will pay our prices."

"Oh, fuck it," said Claude. "I've been waiting for this day all my life." He shook his head from side to side, "Wait until my wife hears we're rich. Oh my fuck, we're really going to be rich."

John stopped him cold. "Nobody hears, nobody. If I find out any one person has said a word to anyone else, I'll kill him."

I looked up at John, he was staring at us all. *I believe you*, I thought.

"Ok," he continued. "You two get back to your jobs. From tomorrow, each of you will carry one bag of specimens off the property. For doing only that you will be paid R500.00 per bag. I will keep six or seven top specimens aside and when everything else has been sold. As long as nobody has talked, you can each select yourselves two of them to keep or sell as you please."

I don't know how John removed the bulk of the specimens from the mine. I know that I carried a bag through the gate every night for four nights running, earning myself a tidy sum aside from the R1100.00 that I later sold my two specimens for.

As for those who found out about the pocket the following week and scoured the level for hidden samples, I can tell you all now, we lowered them down disused ore passes on twine, we hid them in the abandoned tunnel behind the non-white latrine and some were on the bottom of the machine water dam near 3 Shaft. It was at least three weeks before all the rocks were out and in the dealer's hands, but for those three weeks and in the removal of the bigger stones, Basil and I had no role to play.

Solomon came round to my house two days after we found the pocket.

"I hear that you have found a major pocket," he said.

I shook my head. "Not me, Solomon."

"No, no, not you, but your shift. Oh, come on, James, I've seen them. John showed me."

"Well, what do you know," I exclaimed. "The sly bastard. Trust me to lose out."

"Come with me," he said. "Let me show you."

He took me to his car and opened the boot. I looked and there were two large specimens, neither of which came out in my bag.

"They are nice," I said.

He took out a wad of notes as thick as a loaf of bread.

"Why don't you buy for me?" He asked.

He thrust the money at me. "You buy some of the rocks for me as my agent and we can work out a commission. How does that sound to you?"

I shook my head and laughed. "Thanks but no thanks. If I get a rock or two, I will only sell to you, Solomon, but that's as far as it goes."

He shrugged and put the money away. "There are buyers coming in from all over the country," he said. "I live here. You guys think nothing of waking me up in the middle of the night to sell me some specimens. Sometimes I buy rubbish from you just to keep the market going, this costs me a lot of bread. I do it anyway because I am building a business here. I live here, and this is the treatment I get. First good pocket for years and you guys bring in the competition." He looked at me.

"It's no good, Solomon," I said. "It's not my pocket."

"One day, John will want to sell some other stones and he will need me, then I will pay a little less for them." He smiled. "My people have been buying and selling for thousands of years, James. What you are doing here is not good business, take it from me."

He was right of course. John used to get annoyed with the prices he was paid for his stones when we were back to the normal hit and miss, R50.00 here and R20.00 there. Whenever he carried on about that 'fucking Jew', I laughed and reminded him of the 26 Level pocket.

"Yeah," he said, and we would both laugh. "I sure took him for a ride that time."

Claude, true to type, took his money and ran. The last I heard of him, he was shift bossing on a manganese mine. His money obviously had not lasted as long as he had thought it would.

**There is a specimen from this pocket on display amongst the Geological exhibits in the Cape Town Museum of Natural History. It is not one of the most beautiful, but probably one of the largest that we removed, being approximately the size of a large tea tray.

In Pursuit of What?

What was Green Mountain to me?

It was many things.

I first heard of it in Tibet where I was living in a commune with a few other sadly deprived people. We were supposedly seeking a greater wisdom but the price we were paying was too much. If you know deep inside that you lack something, you will grasp at any straw to find an avenue to it. Fortunately for me, I realised that this particular avenue was just that. A dry straw and eventually, I let it go.

One of the people I had met there had a crystal.

It was two blue towers on a dark green rock and for him, it contained a magic that never ended. When we were high, he would carefully unwrap it and gaze into its depths. I too gazed into that stone and, high or straight, its intricacy and beauty would fascinate me for hours. When I decided to leave, he asked where I would go and I asked him, "Where did you get that rock?"

Green Mountain, for me, was a room in the Blue Train. A long line of a building down the length of a back road behind the mine. Here was squalor and degradation, doors broken down, windows boarded up, flies and mosquitoes. The locals called it the Blue Train. Some of the rooms were abandoned, too low even for the dregs of the mining whites. But some of them were a small oasis in a desert of dry holes, with a colourful curtain or a splash of paint to say, "Ok, I live here, but I am better than what surrounds me."

My arrival at the Green Mountain had been marked by a conspicuous absence of everything but myself and that's the way it had been nearly all my life. I did not need possessions; I was happy as I was. The reason for my visit was, however, motivated by that crystal and the purity that had gripped me in Tibet. I wanted to see more and experience more. My whole philosophy was to feel, feel the world, feel sadness, and feel pleasure. Everything had only value in sensation.

Even my innermost thoughts were not mine but only there for me to appreciate. I believed in my soul that we existed as an audience and had no desire to strut on the stage. And that crystal; I desired the wonder it created, the humility it nurtured. I wanted to see more of the place that housed such a miracle, to see more of the miracle itself.

Within a very short space of time, I met a few people and cultivated a relationship or two amongst them. One of these was a master of humanity who called himself Millie. Millie was French and musical; he had lived for years in a commune as I had. He had married, had children and forgotten them. After we had made the usual probes and exposures, we relaxed knowing that we at least were both facing the same direction, if not travelling on the same machinery.

Millie invited me to come around to his place one Sunday. "We can smoke a little grass, drink some Tequila and absorb the world," he said.

When I arrived, he was seated at a table under a flamboya tree; in front of him was leatherwork at which he was carefully busy. I greeted him and sat down. He never said a word but handed me a joint and poured me a shot glass of Tequila. I watched him for a while and then started talking, telling him everything I thought and dreamed. He worked in the shade, the tree above swaying as if in tune with a slow and steady life of which we knew nothing. A dog barked. Someone, somewhere shouted at a child and a Sunday morning sunbath drifted by.

Every now and again as he listened to me, he would say, "Amen brother," but the pile of leather beside him grew steadily.

Much later when the Tequila had seemingly evaporated and a disappointed sun was becoming vindictive, we went inside and he showed me his rock collection. He had no piece larger than a matchbox. He took me to a table and sat me down behind a Heath Robinson collection of tubes and lenses.

Placing a stone in front of this contraption, he said, "This is all you need, man. Check the world in a microscope." He laughed out loud. "Big is not it, small is it. If you go deeper, things get smaller, it's the law. So, I figure if you go smaller, you go deeper."

I looked through the contraption and found myself in a forest of crystals; green fibres shot up into the air beside me, like emerald swords in a fairy battle. All around lay diamonds the size of footballs, scattered as if spilt from a bag on a dark blue carpet. I pulled back and picked the rock up in my hand. It was hardly the size of a marble, the hairs and the quartz barely discernible.

"You see." He smiled slowly. "Deep, man, deep is the answer."

When I went home that night, I took three of his stones with me.

"Can I have these?" I asked.

"No," he gently reminded me, "but you can look after them, instead of me."

In the course of time, I acquired a few specimens and began a collection. I too would get them out from a box under the floorboards and lose myself in their wonder, but they were so few. I wanted to have more and in order to get closer to the source, I made up my mind to work underground. I had very little money and the crumbs from other collections were not enough, no matter how beautiful those crumbs could be.

It was a long slow drag and it was a painful one. I became a learner miner and cast my lot in with the animals. If I had not already led a hard life, I would not have survived. It seems that the gods did everything to exclude me from their treasures. Hard physical labour, ridicule, even physical abuse were experienced and forgotten. Incidents flared and died, and sooner than I had at times believed possible, I was through it and employed in the mining world. I was placed with James McArthur. I had to work with him for three weeks before I could get a contract of my own.

The first thing I asked him was "Do you get many crystals?"

He laughed and asked, "Are you hooked on crystals?"

I shrugged. "No, I'm not hooked, but I have a small collection."

"You will be working with me for three weeks, and I should imagine in that time, we will come across a pocket or two. We will divide everything we find by selection. I will choose first, then you, then me, and so on, ok?"

"That's fine," I said. "I only want some pieces for my collection."

He looked at me and raised an eyebrow. "Not interested in money?"

"No, definitely not," I replied.

There was more than enough for my collection and as people knew where I worked, sometimes a buyer or two would arrive at the Blue Train and I would sell some of the excess I had. In the meantime, my collection grew and I would sit into the early hours and go through it piece by piece. After a while, I bought a small car with the money I made. Sundays I spent with Millie, his uncluttered slide through life feeding me with a little taste of a past that seemed to be slipping away.

One Sunday, he said to me, "Brother, we must rap. Here and now?"

I leant back in my chair and watched the flamboya performing its perpetual dance of obedience.

After a short silence, he started, "The here and now is all there is, because tomorrow is for someone else. You are never the same when time has passed, you know. It's you that's there but, it's not you like you are now this instant."

"That's why I do this every Sunday because it's like the end of the week. I slow down on Sunday. I'm me for a little bit longer. Tomorrow, on Monday, I feel better with the new me because I slowed down and really had a long stay with the old me."

"I want you to know that you are a brother, you feel, man, I have seen you feel, and know you can enjoy. I listen to these guys out there and they are full of shit, man, they are just full of it. They know nothing. They don't have the time to know anything, but they can tell you. Oh yes, they tell you all the time."

He stood up and fetched a fallen flamboya flower from the ground and put it on the table in front of me. "Look at this flower, man," he said. "Just look at it. It is as good as any crystal you ever saw in your fucking life."

I picked it up and looked, but it did little for me. Shaking my head, I told him, "No, Millie, this is not the same, not the same at all."

He took the flower from me and looked at it a long time. "Yes, it is;" he replied. "But you are losing your vision, man, you are going blind."

I felt a twinge of guilt when I left that evening. Millie was a lonely, lonely person and I somehow felt sorry for him.

During the following week, I decided to keep a close eye on McArthur. It was the first week I had worked afternoon shift and I had noticed that McArthur was absent every afternoon for an hour or so at about four or five o'clock. When I casually asked where he had been, he would say, "Oh, I just went down to 27 Level and checked the tramming," or he would name another working place of his.

Somewhere there was something going on and that meant stones. I felt annoyed at being excluded and was determined to force myself in if I could. On the Thursday whilst I was marking off the face, I saw him look around and then disappear through the supports at the back of the working place to the ladderway. I handed the paint to the bossboy and followed. He went down to the sub-level and collected his bag with his chisel and hammer. *I knew it*, I thought excitedly, *it's crystals; he's got a secret cache of crystals.*

From the sub-level, I followed him down to 27 Level and then along the old tunnels in the worked-out stopes to the back end of 6 Shaft. He stopped and watched the 6 Shaft area for some time before emerging and moving over to the shaft itself. In a moment, he had swung off of the concrete onto the emergency ladder and was gone.

I moved over to the shaft and peered down. The steelwork disappeared below me, a sort of Eiffel Tower in reverse. His cap lamp beam was easily seen flickering on the girders as he clambered down. I followed. He went down past 28 and 29 Level and then in a moment of distraction as I slipped and regained my footing, he vanished. Frantically, I increased my pace and when I reached 30 Level, stopped. The mine was deathly quiet; there was no sound, no movement. I looked around, shining my lamp into one eerie cavity after another.

When he spoke, I nearly passed out. I whirled around and he was sitting behind me on a rock, his cap lamp in his hand where he had just switched it on.

"For a man who is taken up with the higher planes of life, you really are rather a suspicious son of a bitch," he said.

"We agreed to share everything by selection," I said hastily.

"What we found in my working places only was what we agreed."

"Where are you going?" I asked. "Do you have some stuff hidden?"

I stopped and taking a breath, continued, "I won't tell a soul, I promise. I will help you and you can just give me a rock or two, good ones of course." I smiled with relief when he handed me the bag.

"You will earn every cent of what you get down here," he told me. "This section of the mine is on a single shift basis all the way to 33 Level. I come down here every day when we work afternoons and check every blast in every working place before the nightshift comes down at seven. It's one hell of a lot of walking. If you want it, you must work for it." So saying, he turned and set off at a wild pace into the level.

Boy, did we walk. Into a level, through each working place, clambering over blasted rock and then down to the next level.

"Why are the fumes so bad?" I asked.

"Oh, they switch off all the ventilation to save power. The fumes clear by themselves if given time but sometimes the bastards blast late and I get the worst of it."

He would rush into a working place, climb up to and examine the face and then shake his head, and on we would go to the next place. When we had finished and had arrived on 33 Level, he stopped and sighed. "Nothing, absolutely nothing. Come on, back to 26 Level."

"How?" I asked foolishly.

"How do you think? Up the fucking shaft, that's how."

And off we went. Level after level, we climbed. I was exhausted. When we were back on 26 Level, I looked at my watch, we had been gone nearly two hours.

"You slowed me down," he said pointedly.

I could only nod in agreement.

The next day, he went on his own and came back with some lead honeycomb. He gave me a small piece but kept nearly all of it for himself.

My piece fetched R200.00.

"Tomorrow I am coming with you," I told him.

"Tomorrow, we have to move," he said. "It's Saturday and we go out early, but you're welcome."

We really got moving with the work when we arrived on shift and at about three pm, I went to look for him.

"Come on, James," I said when I found him. "Let's get going."

He lifted his hand and I saw he was carrying his bag. "I'm ready," he told me.

Then he really travelled as if the devil himself was after him. When he had been in and out of, I don't know how many places, I called to him. "Hey, James, wait up a minute, I can't go on. Let's just have a break, ok?"

He turned and came back down the tunnel.

"If we hit a big pocket, we have to have some time to strip it."

"Oh sure," I replied. "A big pocket? Like yesterday or the day before?"

He shrugged as he sat down. "You never know. This could be the day."

He was right. Two or three places later, as I clambered along a blasted face behind him, I saw a loose rock and pulled it. With a small groan, it slid sideways and I looked into a pocket of dog's tooth calcite on a Duftite base. The fern green balls of Duftite looked like kidneys between which the dog's teeth grew in concentric clusters.

"This is it, James," I called out. "I found a pocket."

Quickly, he returned and we both set feverishly to work. It seemed we had hardly begun and I had only removed one or two pieces, when he looked at his watch and told me, "Only half an hour left."

I stopped in amazement. "You're nuts!" I exclaimed. "We must get as much as we can."

"Stop wasting your breath and dig," he replied.

As I chiselled and levered, I thought, *You're not pushing me around, James McArthur, this is as much my find as yours.*

When he stopped a little later and said, "Ok that's enough, we must start packing the stuff," I was ready for him.

"I have a plan; you go back with what you've got and you keep it all for yourself and I will carry on here. You can finish the work and blast for us both. I will come up just before the end of the shift. Then I will keep what I get."

"Sorry, Gary," he said, "but you are my responsibility and I can't just leave you here. You must come with me." He reached over for the chisel at my feet.

I snatched it up and quickly continued, "Leave the tools. I'll be fine. Look, we will pool the rocks on surface, and I'll share with you fifty-fifty, but just look!"

Desperately, I showed him the pocket, still full of specimens.

"There are still tons of it. There's a fortune here for both of us."

"Sure," he said, getting annoyed, "and who is going to carry tons, tons mind you, of rock up the shaft? Stop fucking around and help me pack this stuff."

I threw down the chisel and picked up a shovel from the ground. I lifted it up in front of me.

"You fuck off," I hissed at him. "You don't even touch one stone here. It's my pocket, I found it. You had already gone past it. So you just fuck off back to the level and leave me alone. I will carry the stuff up the ladder even if you can't."

He was really surprised; he stood up and looked down at me and then at the shovel. After a moment or two, he lifted his boot and slowly brought it down onto one of the better specimens, crushing it forever. "Is this what you are so excited about?" He asked. "Or this?" He crushed another.

I flew at him; I was wild with rage. I came off my feet and went for him, swinging the shovel. He turned and ran.

I ran after him.

He dodged quickly to the left and back to the pocket. Leaning down, he grabbed two more specimens and held them out in front of him like a pair of symbols that he was about to smash together. "Hold it!" He shouted. I stopped and everything stopped. Slowly, I let the shovel sink down.

He walked carefully towards me, holding the crystals in front of him.

"I'll swap you," he said, "for the shovel."

As I put it down on the floor, he kicked it away and gave me the crystals.

"Gary?" He queried. "Are you ok?"

I nodded.

"Whatever we decide about who owns what," he continued, "we have to stop now and pack up. We can't be late and I can't leave you here."

He eyed me cautiously.

"Yes, it's ok," I said. "I'm ok. Let's pack your fucking crystals."

Then I burst into tears.

We got back to 26 Level and neither of us mentioned the rocks. We did not even share them out as we normally did. I didn't have the guts to say anything about them. When we walked out of the gate after the shift, James came over to me and asked if I would come to his house for coffee.

After a half-hearted refusal, I agreed. He still had all the crystals after all.

"You will have to be quiet," he said. "Everybody will be sleeping, but we can divide the stones in the kitchen."

As it turned out, his wife was awake, and I was given supper and coffee.

The incident underground was not mentioned and the stones were divided between us in the normal way.

I thought, sitting there in a warm domestic kitchen, about myself and the other half of my life; the half I had spent in pursuit of inner peace and other noble ideas.

James was a hard and tough man and respected by nastier, more brutal men than I could ever be, but I don't think I had ever been as close to murdering someone in my life as I had been that afternoon. I had no doubt at the time that I was going to do it quite successfully.

On Sunday, I told the whole story to Millie and he asked me how I felt about crystals.

"I want to own the most beautiful crystal in the world," I said and surprised myself with my own honesty.

Millie shook his head. "You can never own anything," he replied. "Only if you know that, can you start to live. Forget all this stuff with McArthur. Next time you get a beautiful stone, hold it in your hand and then give it to somebody. Give it to McArthur."

He stopped and gazed dreamily into the flamboya tree.

"Yes," he said, "that's power, man, that's real power. You have the most beautiful stone in all the world, and you give it to somebody on the street."

I was about to add something but he interrupted me. "You will never do that, Brother, because you're blind; you've got no vision, man."

I left him there and he never asked me to come back. He must be nuts if he thinks I'll give him any specimens. He'll probably just give them away.

Part Three

Majas

Majas is a Fanakalo word meaning 'jacket'. Normally, a thick, heavy jacket. Many of the mine boys are called by this name as they like to be seen in the old army greatcoats. Majas was a short and powerfully squat man who was feared by his fellows and secretly admired by many of the people who supposedly supervised him.

It is easy to say a man is this, or a man is that, but in most cases, men, no matter how simple they seem, are extremely complex. Occasionally, a man is only truly known when the relationship you have with him terminates. The act of severance brings reflections and emotions that finally opens understanding, but too late. Majas did not end his relationship with me at the time of this story, but his involvement ended another association and in the aftermath, I had to realise that I knew very little of the men at all.

When I arrived at the shaft to introduce a communications training programme, the manager, Willem Theron, asked the production mine overseers to each give me ten blacks. Ten from each section meant a total of sixty. The sixty blacks, of varied occupations, would form a team of workers that I would use to replace the labour drawn from the mine for training. The mine overseers, quick to utilise such an opportunity, promptly allocated most of the habitual absentees, the sick, crippled and lazy to my section. Amongst these were dissidents, revolutionaries, and one or two outright gangsters.

Majas was in a class of his own. His current record stated that he had one written warning against him for being drunk on the job. His lapsed record, that is anything over six months old, showed intermittent eruptions of violence that had persisted for a period of some eight years. Despite this, he had risen to a position of gang supervisor and had been one for some time. The change house bull sessions soon revealed a more complex and disturbing history. The shift bosses on the shaft were only too anxious to tell me all about him.

He had arrived on the shaft as a young and intelligent go-getter and, although known to be violent and quick-tempered, soon proved that where he was, the men worked. Consequently, in those less formal times, he had escaped written disciplinary action through the protection of one admiring shift boss after another. He had, however, wasted no time in establishing an aura of fear amongst his fellow workers.

I was told that there was not a black on the shaft that would testify against him. That in the hostel, he was revered as an underworld leader with connections in most of the surrounding black townships. I was told that the number of unproven assaults attributed to him had resulted in his transfer from job to job ending recently in a position as a loco supervisor. In this capacity, he supervised a crew of five, who operated two ore trains. I was advised to get rid of him as soon as possible.

When I first met him, he seemed harmless enough. He was subdued, saying little behind the typical defence system of the mineworker; a shrug of the shoulders, ready agreement to instructions and a point-blank refusal to hazard opinions. He remained that way for several months. His quick understanding and obvious influence over his workers lulled me into believing that his reputation was overplayed.

Oswald, on the other hand, was trouble from the time he stepped into my office; he refused initially to act as a relief, saying that he was already underpaid and had no intention of doing someone else's work whilst that someone was sitting learning rubbish in a training office on surface. Oswald had no qualms about offering opinions. He told me that he had learnt the hard way and other men could do the same. He never changed this self-righteous concept of the situation and was only kept happy by ensuring that he was never asked to do any kind of work except drilling.

His stubborn resistance to me as a supervisor was eased a little by some heavy office politics on my part. I pleaded and gained a re-appraisal of his intelligence rating. Armed with new information, I then demanded recognition of his claim to the position and pay of a driller. Once the personnel department had approved his increase, I asked to see him. He presented himself at my office one afternoon straight after completing the morning shift. He sat down immediately, without being asked, which is a sign of respect amongst the older blacks. He too was short and powerfully built but diminished by age.

I looked at him and he returned the look without blinking. I caught a slight suggestion of brandy and after he had been accorded the normal greetings and respect, I asked him if he had been drinking.

"We all must drink," he replied.

"Yes," I agreed, "but you have come to me from underground, and you are not allowed to drink alcohol when you are underground."

He turned to my clerk, who worked with me in the same office and spoke to him, rapidly and vehemently, in his own language.

I sighed and thought, *Here we go.*

I knew that as soon as he decided to conduct the interview through a third party, it indicated that for him, I had become an untrustworthy person.

Patrick, my clerk, told me, "He asks why have you asked him to come here. Is it to criticise him for his age and his need for medicine? Has the miner complained about his work, because he has done the drilling of two men today. He says this is because the people on whom the mine is spending all this money for training cannot do half the work of an old man such as he."

I turned to Oswald and he gazed blankly back at me.

"Oswald, I have here the papers that say your pay has been increased to that of a driller. From the end of this month, you will receive a great deal more money than you have been getting previously."

He did not even allow his eyes to register a reaction, he turned to Patrick and waited.

Patrick translated for him and he reached over for the slip of paper.

I gave it to him, he glanced at it.

As he gave it back to me, he softened slightly.

"I am an old man. I have sore feet and I need to sleep a great deal; I am grateful for what you have tried to do. At the end of the month when I have been paid, I will return and tell you if it has worked and if I have been given the money."

I wished him well and asked him if there was anything in his life he needed to discuss. Aside from a long explanation as to his need for a drink or two, he was fairly satisfied and eventually left the office.

Patrick told me, "He is an old man and has been working on the mine for a long time. The other blacks say that he cannot drill so well anymore and that the miners get cross with him."

"How do I tell him that he is no good, Patrick?" I asked. "How do I get him to accept anything less than what he believes is his right?"

Patrick was surprised. "It is easy for you. You just tell the personnel officer and they will find him a job as a labourer here on surface or in the compound."

Oswald's position did not improve. Several of the miners he worked for called at the office to complain about his work. I sent my chief instructor to spend some time with him, which did nothing except upset Oswald. So much so, in fact, that he presented himself at the office and berated the inability of all and sundry to handle a machine as he did. He continued with a long criticism of my instructor's attitude and ignorance. This, as he conducted the whole confusing issue through Patrick, took nearly an hour.

The next time I went underground, I decided to visit him where he worked. I entered the section on the top level and was working my way down the centre gully through the working places to where I understood Oswald did most of his drilling. Gilbert, my underground team leader, was with me. We pushed our way through a thick ventilation curtain and came across Oswald.

He was sitting on the ground in the centre of the gully holding his head in his hands. He was covered in mud and dust as if he had been rolling around on the floor. The grey of the dirt was cut by rivulets of blood, which streaked thick, dull, red channels over his neck and shoulders. He swayed from side to side, moaning pitifully, every now and then breaking into a loud cry of despair. Gilbert and I went to him immediately and I squatted down in front of him.

Gilbert tried to move his hands from his head so that he could examine his wounds. Oswald was incoherent and sobbing, he cried out and lapsed into low moaning without making any sense whatsoever.

It was difficult to understand what could have happened. We were alone and Oswald was the only source of information. After closely examining him, we could see that aside from two severe cuts on his head and a large swelling on his shoulder, there was no evidence of serious damage. Gilbert and I raised him to his feet and, supporting him best we could, eventually walked him to the first aid station nearest to us.

We put him down on a stretcher and, together with the attendant, started to clean him up. Once he had been treated, I told Gilbert to stay with him and see that he got to surface safely. I returned to the section to find out what had happened. Nobody knew anything at first. I asked several workers and received the same reaction from all of them.

"Oswald." A shrug. "I saw him earlier going down for another machine."

"Was he alright? Was he drunk?"

"No, he was not drunk, he was fine."

Then I found Corrie Duvenhage, one of the miners who greeted me sarcastically, saying, "Well, well, look who's come down the mine. Are you lost, McArthur?"

"No," I replied, "I came down to see how my boys are doing."

Corrie snorted. "Your boys, you mean all that rubbish you send here to replace my boys." He paused and remembering something said, "Oswald! Where the fuck is Oswald? That kaffir is the most obstinate, useless son of a bitch I have ever seen. I sent him down for a new machine a couple of hours ago and now that I see you, I have just realised, I haven't seen him since."

"I've sent him out of the mine, he's a stretcher case," I said.

Corrie was on the defensive immediately.

"I haven't had an accident in my working places. I've just been through the panels and everything is ok. He's your boy and he must have been hurt whilst hiding from the job."

"Corrie," I interrupted. "I think he was beaten. All I want to know, if you haven't seen him, is where did you send him?"

"I sent him down to the store," he replied. "Look, I'll come with you and we'll ask down on 30 Level."

When we came down the ladderway onto the level, the first thing I saw was Oswald's helmet. I picked it up and looked around. We were standing at the loading area of Corrie's pack transporter. A transporter is a cableway system that carries material up into the working places. Majas was standing on the far side of the tunnel by the transporter's motor.

"What is Majas doing here?" I asked Corrie.

"We had no ore to pull this morning, so I told him to use his crew to transport timber to the panels above us."

A brand-new drilling machine lay discarded on the floor a little way from us.

I called Majas over and asked him, "Majas, why did you refuse to transport that machine for Oswald?"

Without thinking, he replied, "He would not even wait two minutes for us to finish one load; he insisted that we stop and transport his machine." He looked at Corrie and then at me and realised that he had said too much.

"What did you hit him with?" I asked.

He shrugged. "I did not hit him, he fell." He pointed at the ladderway. "He tried to go up the ladderway with the machine and fell."

I knew there was no way that he would revise this stand so I left him and went over to where the other blacks were loading timber. I reached over and switched off the power.

"You people listen to me," I said. They all stopped and stared, sullen, uncooperative. They knew what was coming or thought they did.

I took my notebook from my pocket and waved it in front of them. "Those of you that work for me have their names written in this book, those that work for boss Corrie are known to him. Some of you know me and know that I am a man of truth." One or two of them nodded, and I continued, "I know that Oswald was beaten here this morning and I will give each of you an opportunity to testify as to the truth of what happened."

One man smiled and shook his head, the others were stonily indifferent.

I looked slowly and steadily at each one of them.

"I am bigger than Majas," I told them. "I am going to make him pay for beating one of my men. If I prove that you were here and you helped him by silence or lies, then you will be punished the same as he is. Do not think I cannot do this; you know I have done things on this mine that other men could not do. You know I am the friend of the manager, and you know that if I say something, it is so."

Brave words, but a gamble. If I failed to successfully prosecute Majas, whatever reputation I had built up in the preceding months would be lost. I made sure that I had the company number of every boy that I thought had witnessed the beating.

Corrie Duvenhage was openly sceptical. "They won't testify against Majas," he told me. "He runs the way they live and breathe."

In the course of the week, I called each one of the possible witnesses into my office alone and explained that I would not allow this to go unpunished. I asked each one if they were prepared to tell me what happened. Without exception, they all refused, directly or indirectly. Either they had seen nothing or they had been at the toilet or at the store, anywhere except where they could have seen the incident.

Oswald, thirty-three stitches and a cracked shoulder, the richer was unimpressed by the silence of all concerned and indignantly determined to see

Majas pay for the assault. His version of the incident, told to me directly without the aid of an interpreter, was as follows:

"Mr Corrie told me to change my machine for another one at the store because it was not drilling properly." Here he stopped and then quickly said, "Although I had been drilling faster than the others, I am expected to drill more as I have more experience than them." He continued, "I put the machine on the transporter and I went down the ladder to the bottom. When I reached the bottom, I took the old machine and set out for the store. Majas called to me that I was too old for the job and that you were going to send me home. He said that my machine was fine and that I was just making an excuse for a rest."

He drew himself up proudly. "Majas is a 'tsotse' and I told him that his ways and his loud mouth are the signs of a child whose body has grown and whose mind has remained small. I told him if I were younger, he would be lying at my feet bleeding. Then I left him and went to change my machine. When I returned to the transporter, Majas would not stop it so that I could load my machine. I pushed him aside and switched the machine off."

At this stage, poor Oswald became overcome with the shame of it all and would not look at me in the eye. He mumbled on that he had been thrown to the ground and rolled over onto his front.

"He hit me like a dog, he hit me with a sharp iron like a dog," he said. "Mr James, this man must pay for his lack of respect. I am an old man with seventeen children. I have seen many years and will not keep quiet because the others are scared of a child."

The assault case became of overriding importance to me. In my investigation, I uncovered a file which related the details of an assault case against Majas that was only three months old. It had occurred shortly before he had been transferred to my section and involved an attack of unprecedented viciousness. Majas had apparently gone for one of his workers with a length of cable and slashed the man's face to ribbons.

The personnel officer who had shown me the file said, "Of course, the case was dismissed because there was no case. We could not find a single witness who would testify and the victim himself refused to make a statement."

I was appalled. "This bastard is going through life beating up one person after another and the mine knows it but won't do anything?"

"Not won't," he replied, "can't. Take this case of yours; even if you prove it, we can't fire him. He must have prior offences of this nature before he can be fired."

"What if he gets charged in a court?" I asked.

"Oh yes, if he gets found guilty in a criminal case and must serve a sentence, then obviously he will be dismissed automatically. But you can forget that, no one will testify against him in court." He looked down at the copy of Majas' record that he had prepared for me. "I see that he has a written warning for being drunk on the job. If you can make this case stick, which I doubt, then you will only need one more proven misdemeanour to fire him." He looked up. "That is if you can get all three to occur in a six-month period of course."

I called all the witnesses back one by one and told each of them that I had a watertight case and that it was the last chance that they had to change their minds and free themselves from the possibility of being charged with Majas. To a man they sat in front of my desk, stoney-faced and ignorant. In desperation, I told the last two that I had a sworn statement from one of the others.

Two days later, Majas asked to see me.

"I hit the old man with a piece of flat iron that I found on the ground," he said and stared at me expressionlessly.

I could hardly believe what I was hearing and I quickly asked Patrick to fetch one of the personnel assistants from down the passage as a witness. When he arrived, I asked him to sit down and listen to what Majas had to say.

Majas proceeded to tell me the same story that Oswald had told me and without a murmur signed a statement that I wrote out there and then. When he and the personnel assistant had left, I was happy but puzzled and asked Patrick what he thought had made Majas confess.

"He knows that you will do this thing to him and his spies have told him that you have looked for all the old records. He knows that you will not give up, so he has given up instead. It is a great victory for you."

But I did not believe it and the next day, I called Majas back to the office.

He came in and sat down immediately much as Oswald might have done. His attitude towards me was different, somehow more careful.

"Why did you tell me the truth of this thing?" I asked him.

"You can do nothing to me with the knowledge you have," he replied. "I have a clean record and will be warned. If I am warned." He smiled. "I beat an old man. I am sorry, it is not a thing I do often. I was cross."

"You do not have a clean record," I said. "You already have a warning for drinking."

He waved his hand. "It was a mistake and will not happen again. Mr James, I will not do anything wrong again. My record will stay clean whilst I work for you." He stopped and then continued, "You see, if I do not admit that I beat the old man, then you will try very hard to win this case, and if you do, your name will be very high on this mine and mine will be low. I know that somebody has agreed to testify against me, but I cannot find out which person it is." He smiled.

"Now I confess, I know that you can only spoil my record. Everyone knows that I beat the old man. Everyone will know that the mine can do nothing to me. I will be a good worker; in six months, my record will be clean again." He got up to leave and said, "I see that you have some strength and you will be a good manager. I can help you because I understand the workers and can make sure that your section does well. Lots of the other whites want me to work for them."

I was taken aback by his confidence and became even more determined to remove him from the mine's system. Angrily, I told him, "I don't need you or anybody else to understand the workers. If I am promoted, it will be because of my own efforts and not yours."

He shrugged and left, and I sat thinking about his unassailable position. A criminal charge was only possible if Oswald would agree to go to the South African Police. The problem was that any hint of such an intention would put Oswald in a very dangerous position. I realised that if Majas was found guilty and I had the evidence of signed statements and other documentation from the hearing, Oswald would be safe once he had laid the charge. With other evidence that could not be intimidated, what point was there in eliminating Oswald?

The assault case was scheduled for a Tuesday afternoon. As Oswald had laid the complaint and I was the section head, I chaired the hearing. Tom Brennan was the personnel representative and the other people present included a union representative and a mine translator. Majas had brought two people with him to witness the proceedings, as was his right, but Oswald was alone. Oswald sat through the whole case as if he were a king and only disturbed his dignified mein to give his version of the story.

This he did at great length and in his own language, so that the interpreter was forced to translate his complete testimony. He asked no questions of the others but insisted that every question addressed to him come through the translator.

The affair was cut and dried and I found Majas guilty of assault and gave disciplinary action as a final warning to be entered on his record. Majas was asked if he wished to appeal and he waived that right. I looked across at Oswald who was staring blankly at the other side of the room as if waiting to be sentenced himself. Then I turned and delivered my bombshell.

"I further recommend that the complainant lays a charge of assault with the South African Police based on the evidence noted in this hearing."

Tom put his hand on my arm. "You can't do that," he said urgently.

"Oh yes, I can," I replied. "The hearing is over. He's guilty on his own testimony and has waived his right of appeal. Now he's going to pay."

Majas looked at me, a sneer on his face. He was shocked and angry but somehow still confident.

I turned to Oswald. "Oswald, this case is closed but it is not enough. If Majas is to pay for what he did to you, you must go to the police. I will take you there straight away."

Oswald looked at the interpreter and asked if he could speak. The interpreter nodded and Oswald started a long tirade in his own language. When he had finished, he got to his feet and stood staring once again at the other side of the room. As the interpreter relayed what had been said, I watched him, his old and tired body was rigidly upright but his hands were trembling. I realised that the affair had been a major part of his life and that he was desperately unhappy.

The interpreter spoke as if he was Oswald. "Mr McArthur," Oswald says, "I have worked in a mine for more years than these people have lived. I have seen two of my sons killed digging out the gold for the white man. I came to the mines to get money but I stayed because here on the mines, men used to be men and we worked with our hands. I have sent my sons here to the mines because there are no wars at home and a young man must work instead."

"Now it seems that a mine does not respect age or wisdom but that when one is tired and old, then the young can beat him or spit on him. I am not going to work here anymore; I will return to my family and sit in the sun."

"I wish to say this; I have given much to the mine and I will leave with nothing. Mr James told me that Majas would pay but I see and hear nothing of that. I hear that his record will have a mark on it, I hear that I must go to the police. Will the mine give me a picture of that mark to take home? Will the police force this man to say he made a mistake? If I am to return to my house with at least the respect of an old man, then this mine must tell Majas that he is to say that he was wrong when he hit me like a dog."

"He is to say it here in front of all these people then the mark and the police can be the business of the mine because it is not my business."

We sat in silence and stared at a lonely, old man who had waited weeks for what he thought was an obvious restoration of his damaged self-respect.

Finally, I turned to Majas. "Are you willing to apologise for what happened that day?"

He laughed and said to me, "You want this man to go to the police so that they will lock me up. You want me to say this, go here, go there because it suits your idea of what must be done." He pointed at Oswald. "Your system is not his system and does not help us. We are different; we have our own judgements, our own feelings in here." He leant towards me and pointed at his chest. "Yes, I will tell the old man I was wrong. Not apologise, Mr Royce, admit that I was wrong. You see, you did not even understand that, did you?"

He turned and spoke rapidly at the old man and then left, his representatives trailing him.

I went across to Oswald. "Is it enough?" I asked.

"Yes." He sighed. "It is enough," and then with a surge of spirit, "I am a respected man at my house. This youngster would speak differently if we were at my home."

He officially gave notice that day and two weeks later, was gone. He never came to the office to greet me, and I felt it…I still feel it.

We Are Here!
(Tina Kona)

Three of us were sitting at a split in the tunnel. The main haulage ran from right to left and in front of us the access crosscut for the 62-49 working area disappeared into the darkness, curving away from the well-lit busier tunnel in which we sat.

Thys van Staden was on my left; he was a hard muscular young man, one of the up-and-coming officials. His rugby skills were unquestioned and his attitude told you, 'I am earmarked for seniority'.

I liked him. He was good fun to talk to and we enjoyed a measure of speculation about some of the institutions that other workmates felt were sacrosanct. He called me the philosopher and often asked me to give an opinion on some ridiculous subject such as the ability of a woman to win wars whilst losing all the battles.

Norman Lurie was also a rugby player and had an association with Thys from which I was excluded. Norman was strong but a little short tempered with it and, as he allowed this to colour his relationships with the bosses, not as liable to succeed as Thys. Norman tended to bring conversations down to where he felt comfortable, and we were at this time discussing the chances of the Western Province rugby team in the Curry Cup. As the mine on which we worked was very definitely north of the border, I knew that this analysis could only come to the inevitable and well-supported conclusion that they stood no chance at all.

Whilst the discussion was heating up and also distancing itself from my rudimentary knowledge of rugby, Neels Taljaardt arrived and, without hesitation, joined battle adding his learnt judgments to the issue. I withdrew from the conversation and listened. The knowledge and experience that was professed, even boasted of, seemed to me wasted in this dirty grey hole in the rock. These three obviously should have been busy selecting the teams for the clubs in question, rather than supposedly keeping the gold industry on its feet.

Neels, although accepted in this conversation, was a miner and as such not entitled to become too familiar with those above him. He worked for me and should have been installing tracks in a section of the mine that had deteriorated beyond repair. It had required a complete refit for several hundred metres. After I had given him a reasonable shot at the conversation, I looked at my watch and said, "You will have to do more than sit on your arse and talk rugby, if you want to finish 45 crosscut North by Friday."

He rose from his haunches and dusted his hands, but before he could open his mouth to speak, the tunnel, for one instant, dropped and then seemed to catch itself.

At this moment of incredible movement in our granite world, a flat, harsh 'crack' resounded sharp and resonant, deep in the rock itself.

Norman was on his feet, scrambling for shelter almost as it happened. I, slower than the others, stayed where I was.

The first to speak was Thys. "Shit, that one was close."

"Not just close," Norman replied, "that was here."

I got to my feet and looked down the tunnels into the darkness. "We had better just sit tight for a few minutes and see if anyone comes out from inside."

Norman was immediately argumentative and wanted to leave to check his working places, but Thys stopped him. "James is right. We stay here for a few minutes. If anything happens on this level, it will happen in this area. Anyone coming for help must come through this connection. If nobody pitches up, then we can split up and check our own working places."

Within minutes, our worst fears were realised. We heard the solid crunch, crunch of his boots before we saw his light flickering around the curve of 49 crosscut. By the time he came running towards us, we were already 10 metres inside the tunnel heading in his direction. Nobody runs underground and the rapid fall of his feet told us all we needed to know.

He was a black, who, I don't know. He ran up to us and pointing back down the tunnel, started shouting desperately. "It's a big accident, a big accident. It has fallen badly; the face is closed and the people inside cannot be heard."

Thys grabbed him and held him. "Hey slowly, slowly, we want to help you but you must stop. Tell us properly, where has it fallen?"

I opened my first aid pouch and stripped a dressing from its sealed wrapper. Norman already had one in his hand and the two of us started to clean some of the dirt from his blood-streaked arms and shoulders. Norman picked some slivers of razor-sharp rock from his body.

"This is dyke, that fucking dyke at 16a has bumped."

The black nodded excitedly. "Ya, bass, the small face at number 16 has fallen, but we can hear nobody inside. Long One is digging but I came for help."

"Who is Long One?" I asked.

"He is the team leader," Norman said. Norman had worked these panels recently but had been taken off, replaced by a 'star'.

A 'star' is someone who is the mine overseer's favourite, supposedly able to rectify desperate situations and cause production figures to soar where lesser men cannot. This is quite often achieved by ensuring that regulations and safety are shelved in favour of extra tons and longer blasts. All of us knew that this particular 'star' had sneaked out of the mine early today and was not available. Any action here would be taken by us.

Norman stopped cleaning the cuts on his side and asked, "Is he ok that side?"

"Yes," I replied, "just scratches."

Thys eased his grip on the man and said, "If it is dyke and it has burst, then the whole area will be covered in the stuff; it's like fucking razor blades."

I turned to Neels. "Take this guy to the store and hand him over to the dressing station, he will be ok. Then phone the mine overseer and tell him that we have had a bump at 62-49-16a and that an unknown number of people are trapped. Stay there and wait for anyone else to arrive with any other problems. You know where we are going and you can act as co-ordinator. After you have spoken to the mine overseer, send the store boy for a loco and some flat cars. Tell him to bring the loco to the third, that is the last travelling way in the crosscut."

I looked at Norman. "We will need the cars to transport stretchers. Is that the best travelling way?" He nodded.

"Right," I said. "Hurry, and remember, use anybody you can find to run messages to us or anywhere else, but stay by that phone."

Neels turned to go and then stopped and asked, "What if the mine overseer is not there?"

Thys was abrupt. "Then phone any fucker, man, phone the shaft clerk or the manager's office or the fucking change house bossboy; just get somebody up there to send us stretchers and help."

As Neels took the black's arm, we turned and set off for the inside. My heart was pounding and I could feel that old familiar dread sitting down in my bowels. It does not matter how many false alarms or brutal truths you anticipate, experience and recover from; the bite and poison are never diminished. The sudden strike and the insidious injection of fear into the blood turn on the heat and steal the future.

We passed the first travelling way, dark and deserted. Once a busy access to the panels above, it was now abandoned, the gold long since mined out. When we came to the second one, Thys turned from the track and started up the ladder.

"I'll go up here and check that these crews are ok, then I'll go down the centre gully until I get to 16a. That way, I'll be able to tell you what the fall is like from the other side. If I can get through," he added.

The places were hot and the crosscut was in poor condition. Long stretches of rail track were underwater. It was difficult to move quickly, either the mud was holding your boots back or you had to feel your way through the grey water, knowing that if you put a foot wrong a sprained ankle or other injuries could complicate everything.

We turned into the travelling way cubby and started up the ladder. The way led over slabs of broken rock and up short but steep inclines. By the time we arrived, we were short of breath and tired. The centre gully runs down the middle of the working faces which are mined either side of it. Each working face has a service gully, which connects it to the centre gully. As the face advances, so it moves away from the centre gully and the service gully, called an ASG, gets longer.

This particular ASG had its working face above it on the left and a small extra working face below it on the right. The smaller face, only about 4 metres deep was being mined in rock known to us as 'dyke', and this 'dyke' had burst, closing the ASG with tons of small, sharp flat rocks about the size of side plates. Larger slabs of rock, which had fallen after the burst, had effectively sealed off the ASG entrance with a mass of stone, timber, and broken pipes. Long One was busy with two other blacks; they were moving rocks and debris with their hands.

He turned to Norman with obvious relief.

Norman asked, "What is happening? How many people are inside?"

Long One held up a finger, "There is one, we were drilling holes for a winch bed, the rest of the gang is working on another panel higher up and we had only one man in here."

"What is his name?" I asked and was told 'Mpandlan'*.

I climbed out of the centre gully into the worked-out area and moved down through a small open space towards the other side of the fall. Within a few minutes, I came up against more broken rock.

I tried calling, "Mpandlan, Mpandlan,"* but there was only silence. I crawled back to the others.

Norman looked at me but I shook my head. "Nothing," I said.

"Long One has sent a boy up to the next panel for help and as Thys is coming from that side, we'll have more people here in a minute."

Norman was well in control. Long One recognised his authority from previous times and so did the boys who were helping him. He and I started working to clear the fall in the ASG and passed rocks back to the blacks.

When Thys arrived with four more boys, the work was speeded up. With the sharp nature of the broken rock, care had to be taken and at the same time, we knew that these same razor edges could be the cause of extensive bleeding of the man we believed to be ahead of us. Trapped, silent but we hoped, still alive. Thys was the first to realise that we were close.

He stopped and wiped his face with his sweat rag, not all of us carried a sweat rag but he did, draped in a loop around his neck. I held out my hand and he gave it to me. I wrung it dry and wiped the rivulets of perspiration from my eyes and neck. As I gave it back, he said, "I think we are breaking into a hollow, it seems empty behind these rocks."

He was right, and in minutes, we had made a small opening in the top left corner of the gully next to where the service pipes should have been if everything was normal. Thys crawled halfway through the hole. "I can see him," he yelled.

We waited as he pulled and grunted, his feet scrabbling for purchase, and then waited again after he disappeared.

When he called from inside, Norman was at the hole and waiting.

"Listen," he shouted, "The poor bastard doesn't know what is happening. He is lying at an angle and is buried to his armpits in the dyke fragments. I don't know if he is cut or bleeding or anything but he's alive and keeps saying that he is dying. Get me some water and try to get an air hose connected up and pass it through to me. It's as hot as hell in here!"*

Norman turned and shouted to Long One. "Bring an air hose but only turn it on a little bit." Then he scrambled past me. "I'll get water. I know where the nearest taps are. You get inside and help Thys."

When I had managed to pull myself inside, I looked around. It was hot, I could feel the heat on my skin, and not a whisper of movement of the air was discernible. Two supports and a large slab of rock jammed the roof above us into position. The wooden packs on the sides of the ASG were broken and lay toppled over sideways. We were in a small cave approximately 3 metres wide, 1.5 metres high by 4 metres long.

Thys smiled. "Shit, isn't it?"

I nodded and told him Norman was organising air and water.

He was immediately alarmed. "Only water to drink, we don't want a fucking hose in here. If cold water hits that dyke, it could burst again."

"No, no, just drinking water," I assured him.

"Come closer and help me give this guy some air."

I moved to where he was kneeling beside Mpandlan. Mpandlan sported a heavy beard on a fine long face but as his name implied, was bald.

"Where's his helmet?" I asked.

Thys shook his head. "You use your helmet to fan him and I will try to talk sense to him."

As I fanned him, Thys talked.

The wild starring eyes settled on Thys and locked there. His awareness slowly emerged from some lost depth and when he eventually started to relax; the tension that radiated visibly from him eased. As I knelt and fanned him with the helmet, I listened to Thys, calm and above all, sincere.

"Listen, Mpandlan, it's ok; everything is going to be all right. Tina kona, Tina kona. The first aid is just outside, there by the tip. We are going to dig you out and you will be fine. Now listen, Mpandlan, I am going to feel your face, I will touch you to see if your blood is strong. We are here now and we are going to help you."

Thys slowly put his fingers against Mpandlan's temple.

---

* Mpandlan = Bald one

"Fuck it," he said. "His pulse is a disaster, it's fluttering away at hundred miles an hour but I can hardly feel it."

"Shock," I said tersely. "It could be nerves, but it maybe that he's bleeding heavily underneath here somewhere."

"Wave that fucking helmet," he said.

"You keep talking," I told him and added, "Feel his temperature."

"Ok, Mpandlan, it's all right. Your blood is strong. You are a strong man. We will dig you out and you will be out of here before you know it."

"Where the fuck is Norman with that air hose. This poor shit is going to die in here if we don't do something."

I stopped waving the helmet. "Look, we have a problem here and had better start digging him out of the rock. If he's bleeding, we must find out and quickly."

Thys looked behind him at the dyke. Everything in our little cave was snow white with dust, but the smoother rock of the dyke was evident along the face.

He said, "That dyke can go off again at any time and we might get cut up a bit, but if it does, these few sticks above us are going to collapse and bring the whole fucking roof down. We have only had one bump and normally there are two, sometimes three."

"It has been a long time since the first one; if it was going to go, it would have done it by now," I replied.

We both stopped for a second and looked around at the fragile arrangement that was holding our lives in its chance arrangement of interlocking wood and rock.

"Norman's outside," I said as I heard voices and saw a light flickering through the hole. I went to the hole and Norman's face looked through at me.

"Here is some water." He handed me a plastic jar. "And here is the hose." I passed the water back to Thys and then pulled through about 6 metres of hose.

I brought the end to Mpandlan and played the gentle stream of air over his shoulders and face.

Thys gave me the water. "Here, you look after him and I'll dig."

"Norman!" He yelled. "You open that hole up and we will dig in here."

For a while, I smoothed water onto Mpandlan's face and lips with the sweat rag. Then, for a while, I directed the air at him to cool him off. The very presence of the air in that confined apace was a blessing but it was not enough. Soon, Thys was stopping every few minutes.

"I'm going out," he said. "Norman can give it a go. It will be better if you dig now whilst Norman does that, then he can dig and I will come back in."

"Baas!" The shock of hearing the black speak stopped us. "Baas, my leg is very sore. It is bent behind me and it is paining, but my other leg; my other leg is gone, I cannot feel it."

"It's ok, Mpandlan," I soothed. "We are going to free your legs now, just now. Tina kona, Mpandlan, we are here."

Thys looked pale and were obviously very uncomfortable.

"Get out of here," I said.

He went and whilst he was struggling through the hole, I moved into his position and started to dig around Mpandlan.

I very quickly realised why Thys had stopped digging. Sitting and talking to Mpandlan was one thing, lifting the knife-like rocks from around his body and carefully stacking them so that they would not slide back onto him was something else.

Before Norman was in position, I was already feeling hot, my head was aching and I knew I would not hold out for too long.

Norman talked and I packed. We were down to Mpandlan's stomach when, with my head swimming and muscles trembling, I gave up and crawled from the hole. Thys and Norman had enlarged it a little but because of some very large rocks, had not been able to make it big enough.

Thys poured some water over my head and then disappeared into the blackness.

As I slowly recovered, I looked around and became aware of the blacks, some standing, some sitting, all watching a black hole in the rock.

All waiting, as if some miracle would turn the nightmare off, like a light switch.

"Long One, where are you?" I called.

"I am here," he said simply and took a step forward.

"Get another air hose, we need more air in there. Tell two of these people to fetch it and connect it up, and then turn it on slowly until it comes out fast but not too strong."

He turned to carry out the instruction and I called two of the blacks nearest to me closer. I started them digging on the other side of the ASG, maybe we could get a larger hole if we tried somewhere else.

I decided that we should get some information to Neels so that he could relay it to the people on surface and scrawled the following on a sheet torn from my notebook.

Neels,

Only one black trapped; he is buried and injuries unknown, other places around seem ok. To our knowledge only other injured is the one you took out. Please tell us if news of help. We need stretcher and cold water if poss.

James.

I gave the note to another black. "Take this letter to the white man at the store and then wait for him to tell you what to do. If he sends you back here, bring some cold drinking water. Do you understand?"

"Ya, Baas," he said and taking the letter, disappeared down the travelling way.

When Long One and the two boys returned, we fed the air hose they brought with them through the hole and into the cavity.

Thys called his thanks from inside and I sat back to rest and watch the blacks trying to open up another access. Within a very short space of time, Norman emerged from the hole.

"Fuck it," I said. "I have only just got but here and now I am supposed to go back in."

He was pale and sweating profusely. "Is there any sign of help?"

I pointed to the travelling way where a light was approaching. "Here comes someone now," I said.

It turned out to be 'Spook', one of the other miners on the level.

"Mr McArthur," he said, "Neels sent me."

He stopped and caught his breath. "Whew, that is one hell of a travelling way."

"What is the news?" I asked.

"There is a major fall on 59 level," he panted. "Also on the 49 line. The crews there have really caught it. Two serious injuries and about 10 still missing including the miner. Neels told me to tell you that they are sending most of the assistance to 59 level and that unless you desperately need anything, the mine overseer has said to just try to do your best."

"We have got a possible fatal in there, we need medical assistance," I told him desperately.

Norman interrupted. "Go on, James, get inside and help Thys. I will handle this side for a while."

Spook shrugged helplessly. "You have only one man to worry about and the first aid boy is bringing a stretcher, but it's a difficult travelling way so I thought I would come on ahead and let you know what is happening."

I turned to climb into the hole and heard Norman asking Long One to take some people to help bring up the stretcher.

Inside the hole, the conditions were much hotter than outside but thanks to the second hose, a lot easier than they were originally. Mpandlan was in the centre of a funnel-shaped hole and his waist and the top of one knee clear of the rock. As I moved in towards him, Thys grinned at me, the dust and sweat on his face cracking like the wrinkles of an old man.

"Be careful," he said, "if you touch the wrong place, all this stuff will bury him again…and us!"

The first hose was lying on the rocks whilst the second had been jammed into a crevice. It was fixed in such a way that the stream of air was playing onto the area where Thys sat picking rocks from Mpandlan, lifting them over his lap and packing them into any hole he could find. I took the first hose and aimed it at Mpandlan. Thys shook his head. "He has passed out," he said.

"When?" I asked and crawled closer to the limp but seemingly unscathed Mpandlan.

Thys replied between grunts of effort as he lifted a particularly large rock onto his knee. "Just now when Norman went out. I don't think he has got much time left."

I reached out and took Mpandlan's hand, feeling his wrist for his pulse. It took me three or four tries but I found it eventually, weak and erratic, it confirmed that there was not much time. I turned his head sideways and made sure that he could breathe easily. "Come on," I told Thys, "we must both dig."

We both knew that if we did not get at Mpandlan's legs soon, the man would die. The worsening indications of shock could only mean that he was losing blood and although it could be internal bleeding, we believed that he had been cut by the rocks. The fragments of dyke were so sharp and yet not large, they would cut not crush.

After digging for a few more minutes, Thys stopped me. "Let's try to pull out this leg," he said indicating the top of the knee that was exposed.

I nodded and we both took hold of some trouser leg and at a recognised moment, hauled upwards in unison. Slowly Mpandlan's leg emerged from the rubble and then quite suddenly, his bootless foot pulled clear. A small slide of rubble closed the gap and claimed the boot.

A trickle of dust slowly dribbled from the roof above us, and we stopped breathing for a second.

Nothing happened. We turned back to Mpandlan.

"I can see his other leg," I said. "Here, it's behind his arse."

Thys leant over. "That's his boot too. He's sitting on his fucking leg."

We repositioned ourselves, I moved away and sat down firmly with my back to the hole. Thys moved around to the offside of Mpandlan.

"I will try to pick him up and push him towards you."

"Ok," I replied. "I'll pull on his trousers."

Thys put his arm behind Mpandlan and supporting his head, heaved.

At the same time, I pulled the belt at his waist. Mpandlan flopped over forwards and into my lap. The leg that had been under him was twisted and his foot caught in between two rocks. Thys got his fingers under the one and lifted, straining. Mpandlan came free and as Thys turned to help, I slid out from underneath. Mpandlan lay on his side and we were both able to search his lower body for some sign of bleeding.

It was easy to see; a sticky mass of cloth and dust indicated that his thigh had been injured. I ripped the cloth tearing it away from the flesh and exposed a deep gash about fifteen centimetres long. Working quickly, we turned him onto his back and kneeling over him, I pressed my thumbs into the groin to stop the bleeding. Thys who was watching the rich blood seeping from the cut said, "I think it's stopping."

"Get bandages from my pouch and dress it," I told him.

"I've got my own," he replied as he fumbled at his waist.

When the bandage was on and tight, I released the pressure on the artery and we both watched the bandage to see if the bleeding would resume. It reddened but only slowly and I looked at Thys. "It's ok…Let's get him out of here," I said.

We positioned Mpandlan on his back, head towards the hole and fastened his belt around his chest, under his armpits. I crawled backwards over him going into the hole feet first. When I was nearly through, I took hold of Mpandlan's belt. Thys was standing half crouched over Mpandlan's body and as I pulled, he lifted.

We slid him through the hole bit by bit until the blacks and Norman, fresh and strong, were able to reach him and pull him bodily through and into the open. I took hold of the first water bottle I could see and let it trickle onto my head and down my neck. It was heaven. Thys came over to me having emerged from the hole behind Mpandlan.

"Hey, give me some of that," he said.

I grinned at him and told him to find his own fucking water.

"If I had not done all the work in there whilst you sat on your arse, I'd have the energy to belt you around your ear," he said.

"That's the trouble with the Afrikaner," I told him. "They always resort to violence when frustrated by their betters."

I upended the bottle over his head, and he gasped. We laughed and sat down on the side of the ASG. Just ahead of us, the first aid boy was covering Mpandlan with a blanket before fastening the straps of the stretcher.

Norman came over to us.

"Do you think he will be ok?" I asked him.

"He is ok," Norman replied. "He has probably lost some blood, but he is breathing and there are no other serious injuries, only that thing on his leg."

After the stretcher had gone, we sat for a while, the three of us. About 4 metres up the centre gully, the remaining blacks stood in a small group.

Norman stood up and asked them, "What is wrong? What are you doing?"

"We are waiting, Baas."

"Waiting, waiting for what? Haven't you got jobs?" One or two started to move off and then they all followed.

Norman came back and sat down again.

"Fucking kaffirs," he said.

Strike

"I swore blind I would never come back to this mine and here I am selling my soul all over again."

I am not the best of sights in a near-naked state and my position, spread all over a garden chair, did nothing to help. There was too much fat and too little healthy colour in my already large and hairy appearance. The chair creaked as I shifted and slid further down into it.

Elsabie, my wife, stood up from the laundry basket and pegged a shirt onto the line.

"You needed a job," she stated flatly without looking around.

"Yes," I replied. "I needed a job, but that was a year ago and I am still here; one smashed car, a physical wreck I used to call a body, a double fatal and a near divorce the richer."

"Yes, you drink too much, but that is up to you. You could change that if you really wanted too, and I'm not going to divorce you, that was last night. It is over now, so just forget I ever said it." She picked up the empty basket and went into the kitchen.

I closed my eyes and felt the early morning heat on my chest and face. My head was numb and not too sure of itself whilst my stomach was very definitely on the warpath. I drank because everybody else drank but I used to believe that I was a big boy and that no one forced me to do anything I didn't want to do.

Money was short; the mining house we lived in was small and neglected, first by whatever previous tenants had been there and now by me. Elsabie was right to be unhappy and after eighteen years of marriage, bound to be fed up with sticking it out for better or worse. Last night, she had had enough. Frustrated with waiting until the small hours of the morning, knowing I was not working late but drinking. When I finally staggered in, she had told me she was going to divorce me.

I had to make a change and stop the rot. I sat and thought about what I had learnt from Jason Small, one of the nightshift people I worked with. He had told me that another mine in the area had recently started a new project and that they were looking for people with mechanised mining experience. The wind suddenly picked up and a little whirlpool swept along the concrete yard, spiralling leaves and dust into the air as it went.

A year ago, I had been a mine overseer on a copper mine in Namibia, South West Africa to us older South Africans. My life had been good. I had taken an active role in sports and served on the committee of both an Arts Club and the local school. But the mine had become less profitable so its American owners had laid people off. As an ex-pat, I had been sacrificed and been forced to leave to come back to the hell holes they call gold mines in South Africa.

The truth was that time and Tsumeb copper mine had taken the edge off my memories. I had forgotten the hypocrisy and the focus on production above personal concerns, above lives, above just about everything. What I had thought I could easily handle was proving too much. My behaviour and indolence were symptoms of my inability to come to terms with an environment that was hostile both physically and spiritually. Circumstances, that despite my previous association with them, made me uncomfortable on a basic level.

Everyday a shift boss goes down the mine and compromises his humanity to get the job done. On surface, after the shift is over, the only thing that counts is whether he has produced or not. A mine that ensures that the underground staff receive a week's warning before a managerial visit, is not helping the shift boss or mine overseer have an easier life. Oh no, they, the bastards, are trying to give the poor swine a chance to ensure that the truth has time to be hidden away.

A manager is protected, his visits to the working places are preceded by careful preparation. He is not allowed to see the reality of the mine, so that in all good conscience, he can continue his demands for greater production and reduced accident figures without having to morally perjure his sensibilities.

But I had no cause to complain, having just been part of a team well protected by this system. I thought back to the two labourers killed in my section recently. My working place was an absolute disaster area and after the accident, it had taken a massive effort on the part of all concerned to make the working conditions fit the requirements of the law. Although not directly involved, I had felt dirtied by the incident and most of all, by the fact that I had lent myself to hiding circumstances that could kill again.

It is disappointing to find out that you are not King Arthur or Sir Lancelot and that the pressures of mining life will always flood out the nobler principles you profess to have.

I realised that I would have to do something soon and the thought came to me that if I could not handle this particular moral dilemma, then I must avoid it. Once I had been told that I could change things in the mines, but I had run away then. Now it was time to do it again. I stopped thinking and groaned out loud; this was easily the worst hangover I had experienced for months.

Later that day, when I felt better, I picked up the phone and spoke to Jason.

"Where is this mechanical division based?" I asked him.

On Monday afternoon, Elsabie and I drove out to 7 Shaft on Consolidated Murchison, the mine where I had been directed to look for the mechanical project. I was looking for something known as RDS: Research and Development Services. This group was responsible for developing mechanised systems in Anglo-American mines. In Namibia, I had spent six years intimately involved in a mechanical mining environment.

Having once experienced the efficiency of large machinery, I had never been able to feel comfortable with the outdated manual techniques still clung to by the gold mines. If I could find myself a position where my experience could be utilised in administration or planning of mining machinery, I could kiss the frustrations of underground gold mining goodbye. Ever the optimist, I could see myself becoming a vital part of a project that would revolutionise the gold industry.

My mechanised experience was a godsend to the new project. When I emerged from the 7 Shaft office block an hour later, I had a new job and was a new man. I got into our car smiling from ear to ear.

"Elsabie," I said, "this is it. Not only do I have the job but it pays more money. I have my own office and I only go underground once or twice a week."

My life had turned around. I had a job as a mechanical equipment controller, which entailed acting as a glorified personnel officer to a team of twenty mechanised machine operators and thirty work-study officers. I had to recruit them, supervise their training, allocate them to their various shifts and tasks as well as handle any queries, grievances, or disciplinary action involving them.

I told Elsabie, "They are all matriculated blacks, most of them speak English and they are keen. The manager tells me that they are very careful about who is selected and I have to screen them for the ability to handle responsibility, initiative and relationships with the harder miners they will be working with."

I continued to vent my enthusiasm on Elsabie and whoever else would listen for several weeks. I resigned my position as a shift boss and moved out of the mine house to a small plot just outside of the town. The future looked extremely good and Elsabie and I became husband and wife along older and happier lines.

Consolidated Murchison was a new mine with modern methods, but the mining people seemed much the same. I did not have my own office but shared one with other administrative staff.

Harold O'Mally was a true blue South African of the old school, big and loud with a heart of gold and a fiery temper which flashed and took flame at the slightest hint of disorder. When stressed, his big hands, thickened and scarred by a long and intensive boxing career, would pick things up and put them down with no reason. Slowly at first and then with increasing range and rapidity as he visibly became more frustrated with whatever part of his life had chosen this particular time to step out of line.

It was an interesting source of speculation as to which item would finally end up in his powerful grip when the dam broke and see it flying at the offender, or if there was no offender, at the nearest wall. He always followed up these explosions by leaping to his feet and marching up and down the office, shouting abuse at the system as he explained why this latest plot was directed personally at him and his efforts to get things organised.

He also left no stone unturned in his numerous attempts to obtain information for the people who phoned his desk at all hours in search of data. Data that our office assimilated and recorded in connection with the mechanised project we were running. Anything you wanted to know about the project, you phoned and asked Harry. He would tell you, if not at once, then as soon as it was humanly possible and sometimes sooner. He was widely respected for his accuracy and detail.

The other member of the team that was based in the administration office was Jacobus van Schalkwyk. 'Jacque' was what everybody called him. He was also a South African and very Afrikaans. These two men will always represent to me, opposing poles of the white South African culture. The one, raw and direct, the other smooth, perfumed and absolutely a gentleman. Jacque was never seen as anything less than immaculate and was as fussily concerned over his personality as he was over his clothes.

His character is best summed up by two opposing incidents on the farm he owned outside of town. The first being the story of how he took a whip to two black women who started tearing at each other in a screaming match. It originated over a bag of sugar and was foolishly staged outside his back door. The other of how he drove 600 kilometres to fetch another black woman, the wife of one of his labourers who found herself in labour pains and wanted the 'boss' to bring her back to his farm so that she could give birth in her own hut.

"These people look upon you as a chief," he explained. "If they have a domestic squabble and come to you, because you are the boss, they might say, 'Boss, my husband is sleeping with another woman'." You as a white man think; "Oh for fuck's sake, this is not my business," but you cannot say that.

"For them, it is your business. You must call the husband and ask him 'Are you sleeping with another woman?' and then ask him, 'Why are you sleeping with another woman?' Then you have to say, 'Jonas, if you sleep with this other woman any more, I will beat you…and you, Elsie, if you don't stop drinking and locking your door against Jonas, I will throw you and your children off of my farm. Now get out of here both of you and don't let me have any more of this nonsense'."

"That is what they want. Then you are a person to be respected. If they deliberately show you disrespect, you can only regain it by beating them. They don't understand cleverness or kindness; those things do not represent food on the table or power to them. Their world is physical and they are most comfortable with the old ways."

"What about these people that work for us?" I asked him. "They are educated blacks. They understand about intelligence and education, they are not comfortable with the old ways; they want television and microwave ovens."

He frowned and looked down at his fingernails, splaying his hand out in front of him.

"I wouldn't have one of these bastards on my farm if you paid me. I don't trust anyone of them," he replied.

The three of us inhabited a single office housed in the furthermost corner of the mine's administration block. It was there, before I went underground, before I had a chance to really assess my allies, that I met for the first time, the commander in chief of the enemy.

The enemy became defined in time as most of the underground production team. We had three teams on the project. The production team—Leslie Haughton and his mining men who ran the actual mining operation. The engineering team—Basil Kingsley and his fitters who kept the machinery running. And the administration team—myself who recruited and trained the operators, Harry who collected, analysed and compiled the data, and Jacque who purchased and controlled the supplies. The only thing that the engineering and us had in common was the continuous conflict with Leslie Haughton's mining department.

How the mining people viewed their position on the ladder soon became very obvious when one day Leslie Haughton strode into the office and, hands in pockets, stopped before Jacque's desk.

"You've had mining experience, haven't you?" He stated rather than asked.

"Yes, Sir," replied Jacque. This irritated me because, although in the Afrikaans language 'Meneer' is an acceptable sign of respect to one's superior, the English equivalent in use on the mines is 'Sir', which is not acceptable. Sir is a title with a little more than just a higher rung on the employment ladder behind it. However, no one in the office said anything and Jacque's 'Yes, Sir' hung in the air.

"Good, then you can go home now and come to work this evening. I have a shift boss sick and you can take his place starting tonight. You will go underground at nine-fifteen."

Haughton stared at Jacque waiting for a reply and when he did not get one, turned to leave.

Harry got to his feet. "Wait, Mr Haughton, you have not met James McArthur; he is the new equipment controller."

I also stood up and put out my hand, which was ignored.

Haughton had washed out green eyes and a chubby round face with dark complexion and pitch-black hair. His whole attitude was aggressive and I could feel waves of intimidation being sent out pointedly in my direction. I put my hand down and waited.

"So, you're the expert," he said.

I smiled and shrugged.

"We have had plenty of people here from the copper mines; they never last."

He turned back to Harry and said, "Welkom wants this month's report on time so make sure they get it."

Harry picked up a pencil and put it down. "I sent it yesterday," he started and picked up a book.

Haughton smiled and left. Harry looked at the book in his hand and slowly put it down. "He thinks he runs this section," he said.

"Who is he?" I asked.

Jacque cut in before Harry could reply.

"He is the mine overseer of the production section but he thinks that means he can walk on water and rearrange the universe. He can go and screw himself if he thinks I'm going to work for him. I'm not going underground; I will resign before I do it. I am not employed as a shift boss and nobody can make me do it, not even high and bloody mighty Leslie Haughton."

"You better go and speak to George and explain what has happened," Harry told Jacque. He looked at me. "We fall under George and don't report to Haughton."

Jacque got up and left the office, checking himself in the mirror before opening the door. As it turned out, Jacque did not go underground and Haughton had to find someone else, but that did not stop him from believing that our office was at his beck and call.

Training a front-end loader driver is not an easy task and is, therefore, glossed over by most people who find themselves responsible for it. The recognised method to structure a mechanised training department is to find a misfit at about foreman level and remove him from the production environment where he is proving profitless. You send him on a couple of courses which are carefully designed to expose him to the necessary information but not to test his ability in any way and then you appoint him 'Training Officer'.

You allow him to select a couple of top-class black operators from the ranks, who become instructors. These instructors are the real trainers and they teach the operators to work the machines as best they can. The course contents evolve over a period of time because when anything goes wrong, the operator says, "I didn't know!"

The inspector, or investigating officer, asks, "Why doesn't he know?"

The manager calls in the training officer and says, "In future, teach him!"

The training officer gets hold of his instructors and relays the message and, before you know it, the course includes pre-use checks, some theory on hydraulics and other pertinent details. The courses are written by anybody and everybody, taught by the instructors and bring glory to the training officer who does not actually have anything to do with it.

An analysis of the task the poor operator must perform reveals an almost impossible transformation of skills, from that of a man who cannot even drive a motor car, has never possessed any item worth more than a few Rand and thinks that any education will give him a special place in the world; to that of an operator who can confidently sit behind the controls of a machine weighing over 5 tons and worth in the region of R200,000.00.

To an individual who is circumspect enough to maintain his cool when spat upon by a white, who is obviously his inferior in all respects. To someone who has the strength of character to refuse to use the machine in a way that could damage its components when violently instructed to do so by that same white.

This wonderman is expected to operate the machine at production speed in closely confined spaces, sometimes with no more than a hair's breadth clearance on either side. He is expected to load in and out of tight corners, under large and awkward columns of air and water without damaging either his surroundings or his machine.

This almost miraculous transformation is a regular feature of the mechanised mining world and is accepted by most production supervisors as their God-given right to expect.

Fortunately, this task was one I was familiar with and I lost no time in arranging to meet the instructors supposedly working for me. There were two listed on the gang register but only one attended the meeting; his name was Michael Mkosana. I interviewed him and found him friendly and willing to assist me. He was an extremely large man and when pleased, would lean a little backwards seeming to almost double his height. He was always happy and his smile reflected a genuine pleasure in his surroundings.

I received a letter from him recently and he referred to his well-being in the following way: *According health, I am kicking the highest point and bonny in figure.* This rather quaint self-portrait is as accurate as it is original and I salute his invincible good cheer. On this particular day, his attitude was a little subdued by a question concerning his mining colleague, the other instructor, Peter Dlamini. He shook his head. "Peter is not coming. He says that he works for Mr Haughton, and you must first ask Mr Haughton if you want to see him."

I was not about to expose myself to Haughton's aggressive contempt and replied, "That's ok, Michael, I don't think I really need two instructors anyway. We will just manage by ourselves."

I grilled him about his wages, his housing and anything else I could think of that I could perhaps improve on or increase in order to raise his esteem and position in the section. Once satisfied that I could, in fact, improve his situation in a realistic way, I instructed him to continue as he had been doing until I could find time to go underground and visit him.

As soon as he left the office, I filled in an application to increase his pay rate and to have him considered as a candidate for a course on Basic Instructional Techniques. George Hamilton-Walker, the manager, raised his eyebrows when I asked him to sign both requests.

"What about the other one, what's his name, Peter somebody?"

"I understood he works for Mr Haughton," I said carefully.

"Bullshit," he said and stopped reading to look up at me.

I shrugged. "He sent me a message, Peter that is, to say that I must ask Mr Haughton if I need to see him."

"But that is a load of bloody rubbish. Just hang on a minute and I will phone Haughton, he will tell the bastard to come and report to you."

I stopped him and pointing to the forms, said, "If you sign those, Mr Hamilton-Walker, I am sure he will come and see me sooner or later."

He looked down at the papers on the desk and smiled. He picked up his pen and signed.

"A little kaffir psychology won't do any harm around here," he said. As I turned to leave the office, he stopped me.

"Have you met Mr Haughton?" he asked.

"Yes." I waited.

"Peter has worked for him for years and I wouldn't be surprised if he was also made an instructor as a sort of reward, if you know what I mean."

"Will you refuse to sign any pay increase or other benefits for training staff that don't come from me?" I asked.

"Of course," he replied.

"Then if Peter wants to hitch his wagon to Mr Haughton's train, I'll have to do without him, won't I?"

He smiled again and shrugged. "Best of luck," he said.

Over the next couple of days, I found myself busy interviewing a backlog of applicants wanting work as trainee operators or work-study recorders. Both jobs required top-class people who had a high standard of education. Operators needed to be able to fill in breakdown reports, calculate and record fuel and oil levels. Recorders had to be able to attend and assimilate courses on work-study techniques, keep and record detailed time studies on the tasks the machines performed.

They were issued with expensive stopwatches and special waterproof clipboards. They were trained to observe the work being performed in minute detail.

This very task which was so sought after by the blacks as a prestige occupation was instrumental in making their lives underground a complete misery. Although they did not actually 'spy' on the production team and did not record any information that was not directly related to the machines, they obviously noted aspects of the work that occasionally embarrassed or threatened a miner or shift boss. Without fail, when this occurred, our office would receive reports of the recorders behaving badly, being cheeky or refusing to obey an instruction.

During these first few days of my job, I interviewed and employed three new work-study recorders. All of them from Malawi. They were obviously a group of friends and although, only one of them had any experience, they were all good men. One of them, a shy retiring man in his early thirties, told me that he was a minister of the church and that when he had saved enough money was going to return to his ministry and build a meeting hall for his community. This was a strange concept to me but he assured me that it was common in his country and that three years on the gold mines would finance a hall of comparatively splendid dimensions.

His most striking feature was his English and to listen to him describe his family or country was a pleasure, not because he made it so intimately real but because his accent was precise and his vocabulary so detailed. His education and gentleness were so obvious, that I felt guilty about employing him and wished that I had been hard enough to turn him away. I could not believe that he would retain his faith in so brutal environment as he would find 2000 metres below the surface of the earth. His name was Joseph Khechane.

On the morning I decided to go underground for my first visit, I came in to find two applications on my desk. One was for a pay increase and one a nomination for the instructor's course. Both had been filled in for Peter Dlamini and both had been signed by Leslie Haughton. On them was a note attached with a paper clip: *Please countersign. G. Hamilton-Walker.*

I scrawled *Not approved* in big letters across both of them and put them in a letter addressed to Haughton. I threw it into the mail file and drew a deep breath. By the time I came out of the mine this afternoon, all hell would have broken loose.

I was not to be so fortunate. A few minutes later, Haughton came into the office to speak to Harry. When he had finished, he came over to my desk and looked down at its loud emptiness.

"Did you sign Peter's forms?" He asked.

"No," I said. "I did not. They are in the mail file."

He turned and fetched them. On opening the envelope and reading my comments for himself, he went red in the face. He stopped himself and turned to me.

"Why not?" He asked with a poor attempt at control.

I took a deep breath. "I haven't even met him, how am I supposed to assess the value of a man who refuses to come and see me?"

That did it, what little control had been in evidence, disappeared.

"You fucking little snot nosed bastard, who do you think you are…That boy knows more about machines than you can ever hope to learn. It's only because his skin is black and not white that he isn't sitting in that chair instead of you."

He threw the papers on the desk and stormed to the door flinging it open.

"You will sign those papers and you will bring them to my office in half an hour. Do you hear me? On my fucking desk in half an hour!"

He turned and left slamming the door violently behind him.

Harry laughed out loud and banged his desk. "Welcome to Consolidated Murchison," he said.

Jacque shook his head, his comment was more to the point. "David, what is he going to do to you if you don't sign?"

"He'll back off," I replied, "and Peter Dlamini will either switch camps, which will make me wonder whether I want him, or he will stick with Haughton which will make me wish I had got him."

"He might back off," Jacque replied, "but the shit is going to hit the fan first."

After a few minutes though, I picked up the papers and left the office. When I walked into the production office, I was surprised at the poor state it was in. The mining plans on the walls were old and obviously out of date. The furniture was sparse and rough looking, as if it had been scrounged from a series of second-hand furniture shops. The floor was bare and there was an assortment of old boxes and worn tools pushed against one wall and under a table.

The shift bosses, who were sitting around drinking coffee, fell silent when I walked in.

"Hi. I'm David Royce, where do I find Mr Haughton?"

One of them pointed to a door at my side. "In his office, where do you think?"

I ignored the taunt and opened the door, inside was a smaller replica of the outer office with Haughton sitting behind a large desk, the only other furniture being a filing cabinet topped by some dirty brandy glasses. I deliberately stood just inside the door where both offices could hear me.

"Mr Haughton, may I say something?"

He looked up and nodded, so I continued, "You can tell Peter Dlamini that unless he presents himself to me in my office and agrees to report only to me, I will never sign these papers. As long as he works for you, he can look to you for pay increases and promotion in some other field, but not in training on this section. I am also not prepared to discuss this with you in any way unless Mr Hamilton-Walker is present because quite frankly, Mr Haughton, I am not prepared to sit still whilst you swear at me and abuse me in front of other people."

I turned and left the office as quickly and as with as much dignity as I could. I did not want to give him time to start a slinging match. The shock of what I had said left everyone stunned and I heard nothing behind me at all as I walked hurriedly down the passage. To say I was not scared would have been a lie. When I returned to my desk, I had screwed the forms I was carrying into an unrecognisable mess without being aware of it. I asked Jacque if we could postpone the visit underground for the next day. I had no intention of meeting the enemy on his own ground and losing whatever small gains I had made.

The next day when I did visit the section, I realised that there could never be peace, but that the war between my department and the mining department could only escalate. Jacque had warned me that conditions were bad, and they were. In normal gold mine terms, perhaps really no worse than many other similar gold mines. But for me to see the machines actually operating in these conditions brought home to me very clearly why the gold mines struggled so unsuccessfully to implement mechanisation into their systems.

Everything was dirty; the tunnels were littered with discarded equipment, material, and refuse. The machines were not clean and oil leaks added to the accumulation of dust and grit. The coachwork had never seen paint on most of them and all of them had suffered dents and structural damage of one kind or another.

The miner's rest room was spread along the floor of nearly 10 metres of tunnel and immediately beyond it was the refuelling bay and garage. The refuelling bay, and indeed most of the equipment, had no fire extinguishers. The whole area was a junkyard of nuts, bolts, spare parts and half empty or empty drums of oil. Further into the working area, it was no better. Poor training had resulted in poor drilling habits on the huge drill rigs that had developed this section.

There had been the consequential fracturing of the roof and sidewalls. Huge slabs of granite hung, leering at potential victims as they passed unconcernedly underneath.

After walking through most of the area, Jacque and I returned to the shaft and sat down beside the tip. The tip is a near vertical hole going from one level to another until it exits in the loading level at the bottom of the shaft. All the mined rock gets thrown into this system prior to being loaded and transported up the shaft. Theoretically, if it is empty, the distance from the top to the bottom is measured in hundreds of metres. To fall down, it is certain death and I only know of one incident where a man came out alive, because thanks to a freak chain of events, the one he fell down was nearly full of rocks.

This tip did not even have the normal iron bars forming a half metre grid over it; it was designed to pass larger rocks than normal. It had only two straight rails across it from one side to the other forming a gaping hole nearly a metre wide by 4 metres long. In order to barricade it off and prevent anyone from unwittingly falling into it, a chain had been installed around the perimeter supported on three iron pipes.

Unfortunately, these pipes had been dented and pushed over to such an extent that the chain now hung at about ankle height for most of its length. About three months later, when someone did eventually fall down it, the tip was completely open, the chain having been discarded or stolen.

Jacque pointed to it and said to me. "That is where someone will die, maybe tomorrow or next week but they will die. I have reported it to Haughton and to his shift bosses and they tell me it is going to be fixed tomorrow, but maybe next week. When I was a shift boss, I would never have allowed this kind of work."

I sat back and looked around.

"Jacque," I asked, "what would you do if you were made shift boss in this section now, today?"

"I would refuse, I would not accept this section until I had visited it with Mr Hamilton-Walker and we had made notes on everything here that was wrong. That's number one; then number two, I would stop everything, all the production and fix, fix, fix. Maybe two or three weeks before I would allow anything to happen at the face."

"Hey." I laughed. "I said shift boss, not mine captain. Do you think old Haughton would let you stop everything?"

Jacque became annoyed. "Look at me," he said. "Look at my car, my clothes, my house, my desk. Do you think I would work like this? I say a man will die in this tip and it will happen because these people never do anything until they must; until the mine jumps up and down or until there is an enquiry. So, this tip will only be properly barricaded after someone has died. But I do not do a thing because I care that a black will live or die…I do it because I like things to be right. I wouldn't even eat my lunch in this fucking hole, let alone work here."

"Cool down, Jacque," I said. "I know that but what I am trying to say is that in all my experience on the mines, I have found that nobody up there on surface cares what this tip looks like, or even what the man who will fall in it looks like. What they care about is what it looks like on paper. They care about the reports, the production figures and the enquiries. They care about what it looks like when they see it. They don't give a shit about what is, only about what is presented to them. That way they don't live with what these guys live with that work here."

"They don't live with the Russian roulette of knowing everyday clicks by safely until by some chance the chamber is loaded and 'Bam' your ass is in a sling. Then they all jump back and say, 'but why?', 'But how could you allow that?', 'Why didn't you fix it? Why didn't you tell me?'"

"You see they have the power to screen themselves off and we do not. If you had to work here, you might make it neat but you would never make it right. That is why I am sticking with the kind of work I do now. I feel sorry for these guys; they have to 'please explain' every lost blast, every minor injury report and they have to explain it in such a way that no one can say, 'you are responsible, you are guilty' because too much of that and they get shunted to some other job. I want to tell you that in a gold mine, you must be a miner or a manager, anything in the middle is a dead-end street."

When we came to surface, I went to see George Hamilton-Walker. I knocked and entered his office. "Mr Hamilton-Walker, I was underground this morning and I need to talk to you."

He told me to sit down and listen patiently whilst I outlined how I felt about the abuse of the machines and the immediate need to upgrade the training. With very little trouble, he agreed to let me arrange a comprehensive six-week training course on the afternoon shift, which on this mine was not utilised for production work. After extracting a promise from me to put everything on paper and submit it formally, he asked me what I thought of the mining.

I shrugged. "It's got nothing to do with me."

He was surprised and said so.

"I have plenty of problems," I told him. "When I get more time and my department is running as I would like it to, I'll be able to watch the mining more critically."

He did not push it and I left.

The labour climate on Consolidated Murchison was always in turmoil. A sort of backhander from the aggressive unions in the face of one of the few managements that was willing to be reasonable about the workers' rights and union representation.

Because the mining policy had been modernised to allow the unions to recruit members, hold rallies and establish offices, a sudden flood of minor incidents and grievances surfaced through the offices and workings of the mine.

Skila Patio was a tough, shrewd and well-trained black man who, through persistence and ability, slipped through the security screening and became an employee just before I did. The stories that circulated about him included such items as an unproven murder investigation, training in Moscow and the control of several vice rackets in the area. Few of them were probably true. What was true was that he was mean, totally committed to black power and hated all whites, especially miners, whom he treated with contempt.

His only purpose in finding employment on this mine was to facilitate his obviously prescribed union mission of representing the blacks on Consolidated Murchison. When I was asked to use him in my group, I had already heard of him and seen him in action. It was probably a month before he started to work with me that I attended a disciplinary hearing at which he was present. He was one of several representatives to which any offender is entitled and I was filling in for one of the personnel people that were unable to attend. When the proceedings started, Skila raised his hand and the hearing official asked him what he wanted.

"Who is this new man?" He asked pointing at me.

My name and pedigree as well as the reasons for the replacement were detailed for him. He looked at me the whole time this was being explained as if he was coldly fitting each fact to some quirk of my physical appearance. I was annoyed that this unknown black had the right to query my presence but knew better than to protest.

The hearing did not last too long and very little was achieved. Skila interrupted at every available opportunity with requests for clarity, permission to question a witness and even once to point out to the hearing official that he had made a mistake in procedure. This was carried out with such disdain and frequency that the whole room was irritable and on edge within half an hour. His final stroke was aimed at the miner who had laid the complaint. The unfortunate man, for whom things were becoming extremely frustrating, took out his penknife and started to clean his nails.

Skila raised his hand.

"I would like to direct a question at the complainant."

The hearing official sighed. "Go ahead, Mr Patio."

Skila turned to the miner and asked, "What are you doing, Mr Viljoen?"

"Doing? What am I doing?" Viljoen turned to me. "What does he mean, what am I doing?"

"With your nails, Mr Viljoen." Skila leant forward his eyes as cold and clear as rain. "What are you doing with your nails?"

Viljoen surprised, looked around the room, and said, "I'm cleaning them."

Skila sat up and turned to the hearing official. "I would like to suggest that we postpone these proceedings until the management can instruct its employees to show proper respect for what is happening here."

All hell broke loose. Viljoen leapt to his feet red in the face and penknife still in his hand. Mouthing obscenities, he was halfway across the room before I could get in between him and Skila Patio.

"Sit down! Sit down!" I yelled at him, but it was too late.

Skila was already establishing the fact that Viljoen had a knife in his hand. Behind me, I heard him saying, "I would like to bring the attention of all present to the fact that Mr Viljoen tried to attack me with a deadly weapon."

Later, I heard that Viljoen had decided to drop the charges against the accused and everybody involved was more than happy to wash their hands of an already complicated case.

I had no illusions about what it meant to accept him into the ranks of the work-study recorders.

There was conflict between us from the very beginning. When Hamilton-Walker called me to his office and explained that the management needed to place Skila in an area where he could be less disruptive than on one of the main levels, I tried to argue against it. The theory put forward was that as most of my recorders were educated, they would not be as susceptible to suggestions of rash action as some of the more volatile and primitive labourers. The other advantage lay in the fact that a work-study recorder was as tied to his job as tightly as his watch clicked away the stops and starts of his study.

He could not easily absent himself for the short periods that he needed to incite or guide any work stoppage or aggressive feelings. In the end, I had to accept his posting but submitted a written report to the management deploring the allocation of a known and trained dissident to the section when I had little schooling in the handling of such people.

In a very short space of time, Skila made his mark and the whole section knew he was around. His influence was evident in the increased number of grievances that were formally laid in all three of our departments. Disciplinary hearings, however, became almost a thing of the past once the shift bosses and miners realised that Skila was an efficient and deadly council for the defence in every case. The feeling and atmosphere of the mine were one of irritated expectancy, everyone on edge and overreactive to any sign of disrespect or ridicule. One of the unfortunate victims of this period was my friend, the minister.

One afternoon at about three pm, the three Malawi recorders asked to see me. Joseph, the minister, was the spokesman.

"Mr Royce, we would like to bring persecution to your attention. Persecution of ourselves and some of our colleagues. Not because we are black, no; but because we are educated."

I started to interrupt but Joseph held up his hand and continued, "The whites on our level, those that work for Mr Haughton, resent the fact that we can speak English. This is very evident in their treatment of us. We are not allowed to sit on the benches at the shaft but have to stand at the back when we wait for the cages. This morning, the miner they call 'Hakkies', poured oil all over my jacket that was hanging at the waiting place."

"How do you know that it was Hakkies?" I asked David.

One of the other two said, "I saw him, Mr Royce. I was fetching water with Paulus and we both saw him."

I turned to David. "Are you prepared to testify that you saw him?"

David looked down at his feet. "We don't want to cause any trouble," he muttered.

"And Skila?" I suggested. "If you told Skila, would he want to cause any trouble? I can do nothing if you won't testify, but if Skila represents you, won't that protect you?"

Joseph smiled and shook his head. "Skila is a man who is not here to work. If he is persecuted or even fired, it is an achievement, it is better than wages. For him, trouble is a reward, it is a sign of success."

"Dr Banda, our Prime Minister, does not see things in that light; if we have a bad record or get fired, we cannot get a permit to return to South Africa. Dr Banda is concerned about Malawi and does not want to upset the South Africans. For us, trouble is a sentence of poverty, to be fired from one mine, the end of all mines."

He shook his head again.

"We cannot get involved in unions or strikes; we must be quiet."

I looked at the three of them. "Alright, leave it with me," I said. "I will do what I can."

Shit, I thought and set off for Leslie Haughton's den of lions.

Everybody knew that if I came into the shift boss' office, the chances were small that I was there to pass the time of day. The normal treatment was put into action and I was totally ignored.

"Who does Hakkies work for?" I asked the room at large. No one answered and I stood waiting for a few minutes. Eventually, I walked over to Hendrik Brice, one of the more reasonable of the crew.

"Hennie, who is Hakkies shift boss?"

"I am," he replied. "Have you some sort of problem?"

"Hakkies poured a can of oil on one of our boy's jackets and he was seen," I told him.

Immediately, he turned in his chair. "Who saw him?" He asked.

"One of your people and one of mine," I said.

"Shit!" He threw his pen onto his desk and got up. "Are you sure he was seen? Which one of my boys saw him?"

I shook my head. "Joseph won't say but he assures me that he has two witnesses."

"Joseph is that fucking Malawi, isn't he?"

I nodded.

One of the other shift bosses, James, got up and came over.

"Just hang on a minute, Hennie, he's lying. Your boys are all Xhosa's and Basutos, they won't testify against Hakkies for some pissarsed Malawi."

I interrupted and said quickly, "It doesn't matter, about testifying, I'm not making a case, I just want him stopped."

James sneered and said, "Well, that is a different story altogether, we will stop him. We don't agree with people pouring oil on the poor work-study boys. I will speak to Hakkies myself. I'll tell him to set the fuckers alight instead."

"If you want me to make a case, I will do it," I replied. "I came here to sort it out without resorting to making it official."

I turned to leave but Hennie called me back.

"Just hold it, Royce. I don't like those recorders of yours either, but I will speak to Hakkies. We don't need any more trouble around here."

"That is all I want," I said and walked to the door. As I left, I turned to look at the two of them.

James was watching me and when he saw me turn, he said, "Half the trouble on this mine comes from your recorders, people like that bastard Skila Patio. Do you know he's been trained in Moscow?"

"I heard," I said.

"Well, some of the others are just as bad. This Joseph is a sly fucker, you cannot understand half of what he says. They all sit together in little groups and talk about us. Yesterday, I went over and asked what they were talking about and this Joseph tells me they are praying for a sick friend. Can you believe that?" He turned to Hennie. "This fucking little heathen sits there and tells me they are praying. Who does he think is going to believe him?"

"He is a minister," I said simply.

"Well, what is he doing in the fucking mine? Answer me that, smartass, what is he doing down there in all the muck and shit?" He paused and continued, "I hate the black bastards and one day when you know them like I do, you will hate them too. That Joseph is one of the bad ones and you will find out when it is too late. If I had my way, I would shoot the fucking lot, but I would shoot him first. One minute he is stark naked in the bush with snot running down his nose and the next minute, he is praying for a sick friend in a high falutin' fucking accent."

A week later, James came into our office with some papers in his hand. He came over to my desk and put them dawn, and said, "I told you that I would catch that bastard sooner or later."

I looked at the papers; they were the duplicate forms for charging someone with a breach of regulations.

Joseph Khechane was charged with "poor work performance, sleeping underground and using abusive language."

I looked up and threw the papers back. "I don't believe it!" I said.

"You will have to fucking believe it," he replied. "I have witnesses. Number one, he arrives without his stopwatch and his notepad. Then, when he thinks we are all working, he sneaks off to the shaft and goes to sleep."

I interrupted and asked, "When did this happen?"

"Last night, we are on nightshift this week, but wait, I want to finish telling you. I come to the shaft to phone the office and there is your precious fucking minister, fast asleep, so I wake him up and he tells me to fuck off!"

When James said that I knew he was lying, he saw straight away that I did not believe him and started laughing. "Hey," he shouted. "Rooineck, your little friend is in the shit, and I told you he would turn out bad."

He shrugged. "If you don't want to believe it, so what, you're a cunt anyway."

Before I could reply, Harry's five fingered dinner plates came down on James shoulders and started pressing him down. He struggled but it was hopeless, his legs gave in and he crashed to his knees in front of my desk.

"For fuck's sake, Harry," he moaned. "You are hurting me."

"If you call anybody in this office names like that again, I will do more than hurt you."

"Ok! Ok! I'm sorry, but Royce obviously doesn't believe me and yet I caught that bastard red-handed."

Harry let him go and he left with his complaint forms.

"I just cannot swallow that Joseph would behave like that," I told Harry.

Harry shrugged. "James has to prove it first," he replied and went back to his desk.

Jacque leant back in his chair. "Listen, David, you are inclined to be a little soft in the head about these precious blacks of yours. Maybe you should prepare yourself for the worst, it won't be the first time a white man has been led down the garden path."

I could not accept it and told Jacque, "If it had been Mohau or Atwell, or any of the others, I would say yes but not Joseph. I'll stake my job on it. He might have fallen asleep, or left his papers in the hostel, but he would not have sworn at James like that."

Jacque got the 'well you'll have to learn the hard way' look on his face and went back to his work. I looked across at Harry who smiled and also busied himself with some papers.

"I am going to find out for myself," I said and left the office.

I walked across the footbridge to the hostel gates outside the mine and asked the security to fetch Joseph Khechane from room 216.

When he arrived, he looked most unhappy and I asked him if he knew that he was to be charged.

"Yea," he said, "I know, but that is nothing. I am innocent. I am sick because of what Mr James is doing just to make me feel bad."

"What are you trying to say, Joseph?"

"Mr Royce, last night when we went underground, I left my stopwatch in my coat pocket and after we had left the shaft, I had to turn back to fetch it. When I went back inside, I was far behind the others but I passed Mr James who was relieving himself in the drain. A bit further on are the toilets and I stopped to use them. I put my clipboard and stopwatch on a rock by the door. When I came out, they were gone, but the fitter and his aide were in the store on the other side of the tunnel and they told me that Mr James had come by and picked them up."

"I thought that he had perhaps taken them because he was worried about them. When I went into the working place, before I could ask him anything, he started shouting at me and told me that he knew I didn't want to work, that I was a terrorist."

He stopped and shook his head.

"He was so cross that I decided to say nothing in case he hit me. Afterwards, he told me to go to the shaft and not to come into his section again."

I asked him what else had happened if he had perhaps seen James Harmse again.

"Yes," he continued. "About three o'clock in the morning, he came to the shaft with his pikanin and asked me if I had been sleeping. I told him, no, I had not been sleeping and then he left me alone. When we came out of the mine this morning, he told me he was going to charge me."

He looked down, turning his hands up helplessly. "I don't know why he hates me, but I think you must transfer me to another section because one day, he will kill me or something even worse."

I laughed. "No, Joseph," I said. "He can't do worse than this, in fact, I think he has gone just a bit too far already. I will transfer you to Mr Brice's section, but first, we will settle this complaint."

I was exultant. How I had allowed myself to get caught up in the petty war between production and work-study, or was it black and white, I don't know, but James had so obviously wanted to make me part of his vendetta that I was only too pleased to be able to participate now that there was ammunition on my side.

That night, I was on the mine and waiting for the fitters in their office at the start of the shift. Hans van der Walt came in and put his bag down. "What are you doing here?" He asked.

"I want to talk to you," I said. "Were you on shift last night?"

"Yes." He nodded as he walked over to the intercom book.

"I need to know if you saw James pick up one of my recorder notes and stopwatch last night near the toilets."

"Sure, I was getting some filters from the store and that Malawi of yours went into the toilet, the one that is always dressed so neatly. He puts his stuff down outside like they all do…man, we should move that toilet, the fucking thing stinks. Then a few minutes later, along comes James and picks the papers up on his way inside. Why don't you ask James? If the stuff is missing, he probably put it somewhere safe."

I shook my head. "One more question," I said. "Do you think James saw you?"

"No, definitely not. If he had of seen me, he would have greeted me. No, he was in a hurry and just went straight past."

"Now." I took a breath. "James didn't put that stuff in a safe place, he didn't give it to anybody. In fact, he is laying a charge against the Malawi for going underground without it."

Hans was not surprised. "James is a tit. He has got such a thing about those recorders. I knew he would do something stupid sooner or later. Everybody is walking around on eggshells because the buggers are about to strike, and he does something like this."

"Would you sign a statement saying that you saw him taking the stuff?" I waited a minute whilst Hans paged through the book in front of him. "I don't want to nail James," I continued, "but I have to stop this business, black or white or whatever. These production people are trying to make everybody's life a misery. If they cannot find something wrong, then they make it up."

"Have you got a statement with you?" He asked.

"As it happens no but I'll write one out quickly."

"Well, you had better do it now before I change my mind."

The next morning, I went into Hamilton-Walker's office armed with a copy of the statement and a copy of the charge against Joseph.

He could not believe what I told him.

"We have got more bloody trouble on this mine than we can handle, and the stupid bastards pull stunts like this. What does Skila Patio know about this?"

"I am not sure," I replied, "but I do know that Joseph won't want to make an issue of it. The Malawis are not keen on upsetting the applecart."

"What shift is he working now?" He asked.

"He is on nightshift and if you want to see him, he will be in the office any time now, I told him to report to me when he came out from underground."

Hamilton-Walker picked up the papers and read them through again. "Give me half an hour with Leslie Haughton and then bring Joseph to me. Let's see if we can't squash this damn thing before Skila Patio turns it into a national incident."

And that's the way it was.

A frustrated and reluctantly humble James Harmse dropped his charges, admitting that he had not been sworn at, not found Joseph sleeping and had actually thrown the stopwatch and papers into the main tip. Joseph was given the assurance that James would not bother him again. By the time that Skila had heard of the incident, it was already a golden opportunity lost and life went back to normal.

It was soon after this had occurred that the unrest on the mine got worse. Rumours of security actions in the hostel were rife. There were stories of some gangs underground refusing to work and the always sensitive issue of queuing for the cages at the end of the shift became a hotbed of incidents.

Jacque came out of the mine one day with a torn overall and told us a story of ten whites being pushed aside and knocked down by three hundred blacks when the cage doors had opened. He was bitter and condemned the black troublemakers, the poor security arrangements and the stupidity of all involved.

"The fucking bastards were pushing and yelling long before the cages came," he told us. "We went to the onsetter and said to him, don't open the gates when the cages get here, keep them closed until the bastards are back behind the barricades. James tried to force them back behind the barriers but they threw mud at him and he went crazy. He took his helmet off and started hitting those in the front until we pulled him off. That idiot has got a screw loose; they would have killed him."

"I phoned the surface and told them to send down more security but the banksman said he had been told to treat everything as normal. I swore at him and called him names but he would not budge."

Jacque laughed. "He was so offended. Man, can you believe it? I thought the bastards were going to kill us and this arsehole keeps telling me not to use that kind of language. Anyway, the cage arrives and the blacks are all pulling at the gates when I see the onsetter is going to open the cage door. Before we can stop him, the doors are open and the blacks are like a river flowing over us. I got pushed into a corner and just stayed there. James reckons he got in a good kick or two but he looks as though they got in a few more."

Harry swore and piled his trouser leg up to reveal a gun strapped to his ankle.

"I never go anywhere without this," he said, "and if I meet any trouble, I won't ask questions. I will blast first and check afterwards."

I could not believe it. "You carry that thing all the time. What happens if you shoot the wrong person?"

Jacque snorted. "Nowadays, if he is black, he is the right person. If you had seen what I saw today, David, you would have bought yourself a gun too."

Skila came to me with a letter signed by Mine Manager Horace Weitzman requesting that, no not requesting, stating that Skila must attend some union negotiations and would be away from work for two days. This resulted in another demonstration of this man's power. He was absent for the two days and then another two days after that. I had only signed an absent with permission slip for two days and as Skila had been proven absent without permission before; this second offence should have been enough to dismiss him.

I filled in the necessary complaint forms and charged him, with a recommendation for dismissal. When I took them through to Hamilton-Walker, he was dubious and hesitant to approve the case. Later that afternoon, he came back to me and told me that the management wanted the case dropped.

"I thought everybody was so keen to get rid of him?" I protested.

Hamilton-Walker replied, "The problem is that he is busy with some top-level union negotiations and is essential to the process of getting all the unrest settled as quickly as possible. In fact, Weitzman says that it was him who gave permission for the extra two days leave."

I was not prepared to just accept that and said, "And you, Mr Hamilton-Walker, did you give permission? If it is necessary for someone in this department to countersign the first two days, then how come Mr Weitzman can just go over your head and give him the extra days without even telling us."

Hamilton-Walker looked around. Jacque was listening openly but Harry was ignoring everything as usual.

"I have not got time for this, Royce, I'm telling you that Weitzman gave the man the days off and I, as head of this department, approve of what he did."

He turned and left the office.

I looked across at Jacque. "Maybe you are right," I said. "Maybe a sjambok is the answer."

Jacque denied this. "Oh no, not for Skila, my friend. That particular bastard must be put in a deep hole and buried, that one is poison."

I had the satisfaction of one small victory. Peter Dlhamini had capitulated and reluctantly agreed to work for me. He was an instructor in his heart, he told me, but he was also loyal to Mr Haughton because Mr Haughton had been his boss for many years. I was prepared to accept that and he got his raise and nomination for the instructor's course. The course acceptance papers for both Peter and Michael arrived and the following week on the Sunday afternoon, I picked them up at the hostel and drove them to Vereeniging where Consolidated Murchison was affiliated to a central training school.

I dropped them off, both happy with their new status and the prestige of being treated like whites. I was to pick them up on the Friday afternoon three weeks later. They did not realise what was waiting for them at the end of those three weeks; none of us knew. Back at the mine, we became slowly more embroiled in union problems. For us, it was perhaps not such a shock as it was for Peter and Michael.

Before the mine's labour problems became completely dominant in our lives, I had one more confrontation with Skila. Since the time he had been absent and I had tried to charge him, he was constantly trying to irritate me in any way that he could. He would ask me questions concerning instructions that I had given in a manner that hinted at some lack of ability on my part. He would point out any minor error that I had made in the past as a reason for suggesting that I should check something or ask someone else's advice.

None of this was done politely, neither was it done in a way that I could definitely say was insolent, but he expertly scratched away at increasingly sensitive and jaded nerves.

One of the problems that all supervisors live with on the mines is the inter-tribal hatred of which the strongest illustrations can be found in the Xhosa-Basuto relationship. In many black hostels, it has been deemed necessary to separate these two groups and ensure that they reside as far away from each other as possible.

The antagonism flared up amongst our recorders and it was basically aimed at the two senior recorders who were both Xhosas. Skila was also Xhosa, and this combination of power was upsetting the balance of relationships as the majority of the recorders were Basutos. One of the senior recorders was due to go home for an extended period of leave and I had been priming Paulus Ntiso, a Basuto, to fill his post. Paulus was a thin tall man scarred by acne. His particular value lay in a fine sense of detail.

Harry, who was on the receiving end of the work-study reports, agreed with me. "Paulus can push for more attention to the smaller items like bit changing and fuel stops. We can get Paulus to start recording points that we have been missing up to now."

He paused and continued, "Put Skila onto the same shift as Paulus and that will pull his teeth for him."

Jacque interrupted. "If you want to pull Skila's teeth, get somebody bigger than Paulus, otherwise there will be more than teeth pulled and it won't be Skila's."

When the senior recorder went on leave, I called Paulus in about the implications of what I wanted to do.

"It is not a problem," Paulus said. "This is work. I won't interfere with Skila and Skila must just do his work like everybody else."

"Just don't get involved with any union business," I told him. "It's not worth it. If you have any trouble, then leave it be, walk away from it. If you want to tell me about it, come and see me in the office."

"I understand, Mr Royce," he insisted. "I am not so stupid. I will keep everything in order."

Two days later, I was called to Hamilton-Walker's office. "I understand you have promoted Paulus to senior work-study recorder," he said.

"That's right, he is the best qualified and Harry, Jacque and I agree that he was the obvious choice," I replied.

"Well, he is not qualified enough. Some of the other recorders have lodged a complaint and they want him removed."

I asked Hamilton-Walker if I could sit down and explain. When we were seated, I started and went through my reasoning and the advantages of a better balance of power. However, I could feel that the issue had already been decided and that as in the absent without permission problem of Skila's, I was about to be presented with a blank wall.

To hell with you, I thought. *I won't back down on this one.*

"Who is behind the complaint?" I asked.

"Skila," was the answer.

"Well, Mr Hamilton-Walker, in this case Skila is wrong. Before promoting Paulus, Harry, Jacque and I went through all the records and Paulus is the best qualified; only three people have a full Industrial Engineering Certificate and he is one of them. He has three years' service and he has the attention to detail that we need. I deliberately did not make this decision on my own, I referred to both Jacque and Harry and what is more…" I shifted forward over the desk. "I have personally told this man that the job is his. I don't feel that we have the right to reverse the decision now."

He seemed taken aback and said, "I did not know that you would feel so strongly about this, and there are things I was not aware of."

I interrupted him. "I do feel strongly, very strongly that Paulus must stay."

He held up his hand and went on, "Alright, leave it with me and I'll speak to the manager."

I realised then that this had not come through the usual channels, but that Skila had used his contacts with the hierarchy to lodge his objections. I also realised that Skila was not so interested in protecting his own interest as he was in destroying mine.

The next morning, I did not wait to be called but went in to see Hamilton-Walker fairly early. He told me to come in and sit down. He was playing with a pencil and frowned before he spoke. "David, Paulus must not be promoted, Skila has an objection based on good grounds. Apparently, Paulus has a couple of written warnings on his record and this certificate of his is not a Consolidated Murchison Certificate but comes from another mine."

"That is the beauty of it, Mr Hamilton-Walker. Skila and his cronies are right, it is not even a mine certificate, it is a Technicon Industrial Engineering Certificate and is much better qualification than any mine in-house certificate."

Then I plunged in with both feet. "But that is not the issue here, is it, Mr Hamilton-Walker? The issue is that this bloody Skila Patio is holding a gun at the manager's head and you people are bending over backwards because you don't want to upset him. Well, I'm not scared of him and I believe that snivelling around the feet of a man like him is not doing the mine any good."

Hamilton-Walker got to his feet and said tersely, "That is enough, Royce, it is not your bloody place to decide what does this mine any good or not. Now I'm telling you the man is off—demoted—you can find someone else."

"No, he is not," I replied. "The policy on this mine says that you cannot demote a man from a position unless he is found guilty of certain misdemeanours, and this is not the case. What is more, I personally promoted him and therefore, I intend to lodge a formal grievance against the management of the mine for reversing the decision to promote him."

Hamilton-Walker sat down again and shook his head. "I hope you know what you are doing. The management will not look favourably on an action of that nature. I don't think that any official has ever lodged a grievance against the management before."

"Well," I replied, "the system exists, and the blacks use it; maybe it is about time we did too."

"Just think about it first," he asked. "Demoting Paulus will not mean the end of the world, there are other blacks."

"Letting him down is not the way I work, Mr Hamilton-Walker, and I don't have to think about it, I saw this coming. I will have the grievance on your desk in ten minutes."

Harry said I was crazy. "You will be finished on this mine," he told me. "The management will transfer you into a bum job somewhere and you will just die there. There is no room on the mines for rebels who get in the way, especially now with all this trouble going on. Skila is a VIP you know."

"Fuck Skila," I said. "This is a personal thing between Skila and myself. If he decides who promotes who in my section, I might as well step back and let him run things."

I took the grievance form to Hamilton-Walker.

"You know what you are doing?" He asked.

I nodded. "Yes, I know."

"Ok," he replied. "I'll let you know which one of the managers will see you but prepare yourself for a battle."

The manager, Horace Weitzman, was a small, stockily built Hollander known to the blacks as 'Nyetela', which means 'squash'. He wore glasses that did nothing to subdue his critical and penetrating stare. He did not have a lot to say when he spoke to me in the boardroom, where I suspect he had just met with the union officials, and therefore, Skila. He opened by motioning me to a chair and saying, "So, you are David Royce?"

He looked me up and down for several seconds and realised that I was unimpressed. "How much do you know about the general situation that we find ourselves in at the moment?" His question was accompanied by an intense visual assault from ice blue eyes.

"Not too much," I replied. "I know that the mine is having severe labour unrest problems."

"Well, sit still for a minute and because of the rather unusual position that you find yourself in, I will take the time to explain to you what is going on here."

Well, bully for you, I thought.

"Approximately, six months ago the official representatives of the union were nominated here on 7 Shaft and when the management accepted those nominations, the people were duly appointed and we had a union. Initially, we had quite a few grievances but they were expected and represented a minor flexing of muscles, which did us all some good."

"However, that did not last long and we sensed a push in a definite direction, finding that emphasis was being laid in issues, which had either political motivations or were smaller but very sensitive and awkward. In short, publicity orientated. The union wants to grow and it needs fuel to fire up membership on other shafts and mines. The arrival of Skila Patio heralded a new era in the level of negotiations. He has been well trained and is very, very smart."

"He is our local chairman, regional representative for the Western Transvaal and has links with COSATU, which, if you don't know, is a hard and powerful political force in the country and represents all the more dissident unions. We believe that we have been selected at 7 Shaft as the place for a major demonstration of union power."

He hesitated and then got up and stood at the window. "We have to monitor our behaviour on this road we are being forced along very carefully. We must eliminate any superfluous problems as quickly and efficiently as we can so that we can apply our minds to the key issues with as few distractions as possible. Skila is an expert at undermining just that. He knows what he intends to use to carry a major strike and can quite cheerfully throw in three or four other grievances a day which we have to just as quickly deal with."

"People like you play right into his hands and through stubbornness." He turned and held up his hand. "Justified in some cases, prolong the settlement and tie up the personnel team in investigations of records and enquiries over minor situations when they could be more strategically utilised."

He returned to the table and sat down. "We have on this occasion changed our minds and will back your decision to promote this…." He waved his hands at some papers in front of him. "This Paulus. However, as Skila Patio works in your section, I feel that this explanation is warranted. Not that you will see much of him in the near future. Things are coming to a head, and he is going to be heavily involved in some high level negotiations over the next couple of weeks."

"I want you to cooperate with us, don't get involved in any change house bullshit sessions about what you have heard here and don't antagonise Skila by any comment on your little victory. We will inform Skila and Paulus separately through the personnel department of these decisions."

He stopped and looked at me but did not seem so intimidating. "Does that satisfy the issues behind your grievance?" He asked.

"Yea, thank you very much," I replied. "I am quite happy with that. I only wanted to protect Paulus."

"And nail Skila Patio," added Weitzman. "By the way, we have not recorded a formal grievance and I intend to commit these forms to file 13." He threw the grievance forms in a wastepaper basket. "Your action is an unprecedented step on the part of a white official that we do not wish to encourage."

He got to his feet and so did I.

"Thank you, Mr Royce," he said, and I followed him out of the boardroom.

I walked to the office feeling elated over my victory and concerned about what I had learnt. I was glad that my two instructors were well out of it and living the good life at the training centre in Vereeniging.

The following Sunday afternoon, a routine security patrol heard activity in the welding shop at the shaft bank and on investigation, surprised six blacks busy manufacturing weapons. The blacks escaped and lost themselves in the crowd around the hostel beer gardens. The weapons were gathered together and put against the wall outside the manager's office ready for his inspection on Monday morning. When morning came, the weapons were gone, stolen by some unknown person after the Sunday nightshift had gone underground.

Hendrik Brice, who had seen them before going down the mine, told us the next morning that they were for the most part very crude, mostly sharpened rods or club-like pipes. One of them, he told us, was a bicycle cog slotted into the top of a steel water pipe forming a sort of spiky battle-axe. All of them were deadly.

That same Monday, I learnt that Paulus had been seriously beaten up, walking back to the mine from the village shops. He was in hospital and I went to see him. He told me that he was going back home when he recovered. There was not much that I could do about it. He was wise and knew better than I that Skila was an adversary he could not beat.

Harry and Jacque had news for us as well. They had been through to 8 Shaft early in the morning to fetch spares for a major engine overhaul on one of the front-end loaders. At the circle, 2 kilometres from the shaft, was a burnt-out car shell. The store man at the central diesel store had told Harry and Jacque that the car belonged to a Shangaan bossboy who was not prepared to involve himself in the protest march on Saturday afternoon. He had been dragged from his car and covered in petrol. He and his car had provided a spectacle that had fired the emotions of thousands of chanting mineworkers for several hours.

We, that is, Harry, Jacque and I, decided that we should travel to work together in one car, taking turns with the driving each day. Two weeks had passed since I had taken Peter and Michael to Vereeniging and I phoned the training centre to confirm that I must pick them up on the Friday.

On Tuesday, none of the blacks at 8 Shaft went underground, instead, they marched on 7 Shaft and were stopped at the hostel gates. There was a footbridge over the road from the hostel to the shaft so that the blacks had easy access from hostel to shaft and back. We stood on this bridge and watched the scene unfold only a few hundred metres away.

For the security and management, the situation was explosive and volatile but for the blacks, it was ceremonial. They sang and swayed to their own rhythms and music. The leaders and the management glared and gesticulated. Every few minutes, they would break up and return to the major groups, then walk back to each other and glare and gesticulate again. Nearest to us, the armed power of the mine security stood silent and watchful, guns at the ready with their vehicles lined up behind them, a solid wall of khaki green across the road.

Further down, were the smaller groups doing the negotiating and behind them the solid mass of chanting, undulating black power.

The human force is, however, not always as dedicated as its manipulators might wish and the negotiations were rendered irrelevant by the heat and rapidly dissipating energy. The security forces stayed restlessly in position but the mass of blacks slowly filtered away to form small groups under the trees in the surrounding fields. If you looked beyond the major group, you could see people making their way back to 8 Shaft in twos and threes.

We found out afterwards that due to a breakdown in the communications of the union, a uniting of the two labour forces of 8 Shaft and 7 Shaft had not taken place. All of 7 Shaft was underground and unaware that they should have been on surface to meet with their marching brothers in a show of force. The fact that the 8 Shaft people found themselves alone did much to undermine any resolve that existed.

On Tuesday afternoon, we were in Harry's car and on our way home at four pm. We pulled up at a set of traffic lights at the junction of the mine road and the national road that led to the nearest town, Viljoensdorp, where most of the whites on the mine lived. Although the traffic lights had changed against us, the first car at the white line on our right had stalled and was holding back the traffic behind it. The car immediately behind the stalled car was one of the hundreds of black taxis that operated in the area and the driver was impatient, pumping his hooter two or three times.

The driver of the stalled car put his hand in the air showing his finger. We sat and listened to the starter churning and the blacks hooting.

Suddenly, the driver of the stalled car leapt out of his car and shouted at the taxi driver, "You fucking kaffir! Shove that hooter up your ass, you black bastard!" And scooping some gravel from the road, threw it at the taxi windscreen. The taxi erupted and blacks spilt out onto the road. The taxi driver was beside himself with rage and taking his shoe off, he held it above his head and charged the white man followed by his passengers.

At that moment, the lights changed and Harry shot across the intersection with his wheels spinning.

I watched the white driver running down the road away from the intersection. Jacque said, "Look at him run."

"We can't leave him," I said, "those guys will murder him!"

"You cannot afford to get mixed up in these things," Jacque told me, and touching my leg motioned at Harry. Harry was as white as a sheet and his big hands were locked rigidly to the wheel.

We were belting along at twice his normal sedate pace.

Hell, I thought. *He is the one with the gun strapped to his leg.*

On Thursday evening, the hostel manager had to lock himself in the store to avoid the attention of a mob of approximately fifty looters that had started off as a group of three people unhappy about their evening meal. They occupied his office for an hour until the mine security arrived to drive them out and release a very shaken and disturbed man.

The next day was Friday and half of the workforce refused to go underground. We found out that a similar situation existed at 8 Shaft, but that most of the other shafts were operational; the blacks biding their time, content to see what would happen to the more rebellious groups before they committed themselves. I fetched a car from the motor pool and set off for Vereeniging to pick up Peter and Michael, very happy to be out of it for most of the day.

My two instructors had obviously enjoyed themselves and were, in Michael's case, broader, sunnier and expansive, and in Peter's case, smoothly friendly, as if he had seen and been conquered by the glory of his training future.

In the three-hour journey back to the mine, they quizzed me about the events of the last few weeks as they had heard some disturbing rumours and seen rather inadequate reporting in the news media.

Michael shook his great head. "Mr Royce, these people are foolish. Before we left, I told them, to change the world is a fine thing because there is much wrong, but it must be done slowly. We must sit and look at the road for a long time, we must listen to people who can tell us where there are places we can rest and eat. There are many paths to the next village and we should be able to choose which one is best."

Peter, younger and I expected less patient, although I did not know him well, interrupted. "Sometimes it is not enough to just go somewhere. You must walk where you can be seen so that people know you are coming. It is hot out in the sun, the older men need to find shade and drink beer for strength but these people who strike, they are strong and walk in a straight line."

I asked him, "Do you think it is right to burn the Shangaans because they do not agree, or break down the beer gardens because they are run by the whites? What about the mine? If the mine dies, then where are the jobs and where is the money for your families at home?"

I knew that it was possibly dangerous ground to open up and that I could find myself deep in an argument that I was not capable of handling. It was refreshing, however, to hear someone who did not hide behind sullen indifference or false platitudes.

There was a short silence from the back of the car before Peter answered. "Michael and I were the only blacks at the training centre; we sat with the whites, slept in the same sort of rooms, ate food that I have only seen in magazines."

Michael grinned and patted his stomach.

Peter continued, "It was good but it can only be for certain people, not for all. Whilst I was there, I realised if everybody had to share what was there, it would not be enough, but if the people who have it want to keep it, they must be able to do so or they will have it taken from them. Those strikes, I think that the black people are saying, 'watch out, we are coming. We think we are strong enough for more'."

"Are you married?" I asked him.

"Yes, Mr Royce, I have a wife in the Transkei. She is a teacher."

"Are you not worried about losing your income if the mine had to close down."

He laughed and said, "The mines are as old as the hills and the trees, so long as the white needs money, he will need to go down the mine to fetch it. When I first came to a mine and saw how the shaft goes down into the earth like an arrow…how the light is taken in wires to the blackness so far inside…and I saw the many people who work to dig out the rocks that we threw at the jackals on the hills, I knew, as all of us know, that nothing will ever stop the mines."

"You are wrong," I said. "A mine can be stopped."

As we approached Consolidated Murchison, I explained to the two of them that I intended to call at the mine offices to arrange a leave of absence for them for the next day, which was Saturday, and then drop them and their luggage at the hostel before returning the pool car. They were pleased to think that they would have a free day and easily agreed.

At the mine, we were stopped by a roadblock but waved through on production of our security passes.

Overhead a helicopter roared by and then hovered low over the hostel area. I identified it as the helicopter used to ferry the top administration backwards and forwards to Johannesburg and realised that it was being used to observe the situation by our own, probably very worried, executives.

When I arrived at the office, it was empty.

I turned to leave and saw Jacque coming down the passage. "Come on," he said, "get your stuff, we are going home."

"But it's only three o'clock," I protested.

"Well, whilst you were fetching your wonder boys from wherever, things have progressed. All the boys stopped work at ten am this morning and demanded to be pulled out of the mine. We have had a minor riot in the hostel and some blacks got injured…So," he shrugged. "The management have told us all to go home."

"I need to see Hamilton-Walker," I said.

"He's not here. He is at some meeting or other which involves everybody above section manager level."

"Who is the highest ranking official in our section who is available?"

I stopped and raised my hand. "It's ok," I said. "Haughton, right?"

Jacque nodded. "He and the engineering supervisor are in Trelawney's office having a dop."

"Don't leave without me, Jacque. I must tell someone that I am giving Peter and Michael the day off tomorrow and then I must drop them at the hostel before I return the car. Fifteen minutes at the outside, ok?"

"Day off!" He laughed. "They don't need you to have a day off, the whole fucking mine has just taken a day off."

"That is the point," I replied. "It could be important that I have a senior official to witness that I gave them a day off and that I am not covering for them."

I left and went to the engineering supervisor's office. His door was wide open and Harry, Leslie Haughton, Jimmy Trelawney and one of the loss control people were sitting inside. The place was thick with smoke and the brandy bottle was nearly empty. When I went in, Leslie Haughton asked me belligerently, "Where are Peter and Michael?"

I said, "They are both outside in the car. I want to know if they can have tomorrow off?"

He was surprised. "Are you asking me?"

"Mr Hamilton-Walker is not here, and you are next in line."

He was obviously impressed with the thought and became a little more tractable.

He asked me, "Why should they have a day off when they have virtually been on holiday for three weeks?"

Trelawney interrupted. "Give them the day off for fuck's sake, nobody will know the difference; nobody will be here tomorrow to know one way or the other."

"Yeah, what the hell, give the fuckers the day off, Royce." Leslie Haughton, grace itself, poured brandy into a glass as he spoke. He pushed the glass across the desk and said, "Here, Royce, have a drink."

We looked at each other; behind his attitude of live and let live, lay a strange intensity that put my back up.

"No, thanks," I said, "I want to go home. Are you coming with us, Harry?"

Harry explained other arrangements and told me to tell Jacque that we did not need to worry about him. I left the mine and went home.

What happened that night is history, there is possibly a factual and detailed account of it in the mine records. I was not there until later and must rely on what I was told by others.

I was told that of the seven thousand hostel inhabitants at the shaft only two or three hundred actually took the law into their own hands. I was told that when the management decided to fire the whole labour force of 8 Shaft and 7 Shaft, some sixteen thousand people, they wisely did not relay this to the hostel crowd. Instead, they laid on a fleet of buses and told the blacks that it was necessary to clear the hostel areas. They convinced them to cooperate and board the buses.

The more tractable of the workers, worried by the rising violence and the increasing tempo of chants, wild shouts and security action, agreed only too readily. This strategy had been decided on in advance if circumstances met specific conditions. Key personnel had been primed as to how to proceed if the go ahead was given.

The mine's own transport system was ready to take to the road immediately and a stream of buses ferried the compliant blacks from the two hostels to a nearby sports stadium. At the sports stadium, all was ready. The floodlights were on and the security fences and other measures were manned and waiting. Erected and designed to control over-enthusiastic spectators, they were now ideal to regulate bewildered labourers being shepherded back to their homelands.

Soon, despite the attempts of the hardline strikers, the majority of the hostel inhabitants, mostly the Shangaans, Malawi's and supervisors were milling around inside the floodlit security fences of the Oppenheimer Stadium, the mine's flagship sports complex.

When I arrived at the stadium in the small hours of the morning, having been called from my bed by Hamilton-Walker, I assisted in firing and paying out the entire workforce of the two shafts. That night, I was something I would never forget.

At the sports complex, there were nearly ten thousand reasonable but very confused and frightened men. At the hostel, maybe three hundred drunken, rampaging savages intent on destruction.

At the complex, I found shift bosses, mine captains and personnel teams sorting and pacifying thousands of black workers. These workers were bewildered and frightened men clutching what few items of value they had been able to pack in the time they had been given.

Back at the hostel, armed security forces, police advisors and management representatives were trying to round up and remove frenzied and vicious groups of killers. These men were seeking out and tearing down anything that represented the management, reason or discipline.

I spent most of the Saturday at the sports stadium; part of a terrible but seemingly necessary demonstration of the management's right to run a business at a profit. I spent most of Monday in the hostel; part of what started out as a sensible and humanitarian action on the part of the hostel administration but turned into a sickening demonstration of the pettiness and greed of the people I worked with on a daily basis.

On arriving at the stadium, tense and expectant, I was met by security officials who asked where I worked. When I told them, they directed me to one of the gates and told me I would receive further directions inside the stadium.

Inside the stadium, there was a mass of harsh lights. At the northern end, the large gates leading into the centre section consisting of the field and athletic track were open and jammed with security uniforms.

The buses from the hostels were unloading at this gate and the blacks were filtering from them through security to the centre of the field. In the centre of the field were people directing them outwards to the spectator's stands to join their workmates already there. It was interesting that an edifice designed to provide these same blacks with an afternoon's entertainment could so readily fit its new and totally reversed role. The western section of the stands had been allocated to our shaft and the eastern side to 7 Shaft.

The blocks of seats were allocated to the various working sections within the administrative structure of each shaft. When a man arrived in the centre of the field, he was asked, "On which shaft do you work?"

His reply would get him sent to one side of the stadium or the other. He would be met at the gates leading from the arena to the seats and his number checked against a computer sheet, which would detail his gang number and section. With this information, he was ushered to the block of seats allocated to his work group.

The noise was immense, immense because of its non-threatening nature and its rocklike consistency. It continued as a loud backdrop, meaningless and rising high above us. There was so much to that scene, so many aspects to astound and throw open one's senses, that it is perhaps best to initially list impressions that have remained after many years having sifted them.

It was an incredible undertaking on the part of the management. The logistics were staggering, no man was to leave the stadium without his record of service, his contract termination, his full pay, and all his savings and pension monies accrued over years of employment. It had been meticulously planned. Bank staff from all the banks in a 50-kilometre radius of the mine had been wakened and asked to strip their coffers of all available cash. This money was collected and transported to the stadium by a fleet of payroll cars seconded from any security company available.

I have been told that the sum of money paid out in notes and small change at the stadium's ticket boxes was in excess of R7,000,000.00. It required the complete reversal of the normal use of a sports stadium. Normally, a spectator comes to the turnstile, which prevents him entering until his ticket is paid for, and then proceeds to the refreshment kiosks, where he buys something and goes to his seat. There, he watches the spectacle and leaves through the exit.

Here, he arrived at the exits, became the spectacle, then proceeded to the refreshment kiosk allocated and collected his papers. Then he was taken to the turnstiles where he queued to collect his money before boarding a bus for his homeland.

Without any question, the greatest tragedy for all of us was the sudden betrayal of faith that was perpetrated. Every miner, shift boss or department supervisor has a story to tell of how he was approached by his people with confidence, sometimes in groups, sometimes individually. Now that the boss was here, he would explain, he would somehow cut through this mad confusion and return them to sanity. Now they had found the man who worked with them, who knew they were not troublemakers, knew they had always supported him, stood by him when others had tried to stop the work.

Now that they had found the man who had in some cases worked and shed blood with them for years. The man that was different, not like the others, their boss. Now things would be fine! But these gods, themselves betrayed, could do nothing and that night a myth was destroyed for thousands.

The black man does not easily lose faith but once it is lost, it is gone, like a coastal wind it is warm and pleasurable and when over, lost forever to the sea. The lesson that was to be taught would have been harsh, thousands without work, poverty and hardship rife in the homelands, if it had affected the worker only physically, but it was more than that. A blow had been coldly and severely dealt to the dissident unions, their boasts had become meaningless and their power illuminated as a weak thing compared to the unassailable position of the gold industry. Peter had been right the previous day; nothing could stop the mine.

The union had used the labourer once again to throw its fist into the air and scream defiance to the world. But this time, the company had struck back. The management had used the labourer too, sweeping him off the negotiating table with a single stroke. And lessons had been learnt; management and the industry were stronger and knew it was back in control but the cost had been high. The psychology was flawed, migrant workers were used to going back to their homes for extended periods. Essential skills had been lost and it took a long time to get them back.

However, the union was decimated, shop stewards, officials and credibility, all on buses to nowhere. But the people, the labourers and semi-skilled had pensions interrupted, continuous service records carefully protected by wise and thrifty minds, destroyed. Even worse than this, they had learnt that the all-powerful white miner, who ruled their violent underground environment with an iron fist and scorn, was as helpless and as ruled as they were.

It was not the whole industry that was there; it was not even the whole mine, but it was enough. The message was strong and it was heard throughout South Africa.

During the course of the night, I ferried the recorders and equipment operators one by one to the refreshment kiosk nearest us. Each one would tell me he had done nothing wrong, that he had left the mine with the others as his life was at stake. When we arrived at the counter, he would give his number and the miner or shift boss behind the counter would find his papers and pay slips. He was asked to sign for them; if he refused, the official signed a column marked 'Refused to sign'.

He was then led to the nearest turnstile, queued for his cash and put on the appropriate bus. Some of the longer serving men climbed aboard overcrowded buses with few belongings but in excess of R5000.00, sometimes R6000.00. Security for them was a case of concealment; if the money had been seen, it was coveted.

Peter was there, indignant and not prepared to talk to me. He wanted Leslie Haughton, confident that Haughton was big enough to make his case different. Michael arrived and with eternal optimism, informed me that he would turn around at home and be back at the mine in a week.

"Mr Royce, it is simple. I was not here, Peter was not here. We officially worked a full day in Vereeniging and no one can deny that. So, you see, I go home for a week and you see the personnel officer on Monday by the time I return, all is well and we continue as before."

It was actually nearly three months before he was re-employed and it took bitter and protracted fighting to reinstate his and Peter's privileges. I was dreading having to face the three Malawi men who had hardly begun before they were stopped. As it turned out, they were not there, and I did not see them until Wednesday the following week.

Early on Saturday morning, I went to a central canteen erected for the whites; the blacks were fed regularly from mobile kitchens positioned around the arena perimeter. I took some coffee and was walking back to my section when I met Leslie Haughton.

We faced each other on the concrete in front of the pavilion surrounded by still crowded stands and busy officials. I was tired and apprehensive. He was drunk and slightly flushed. "Have you seen Peter, Mr Haughton, he was looking for you?"

"Yes," he said slowly and deliberately. "I have seen him, and I suppose you are bloody pleased with yourself, pleased with the fact that I have lost a fucking good man."

"But we have all lost everybody," I replied.

He went red in the face and started yelling at me.

"You, you slimy little shit, you gave him the day off. Who told you to give him the day off? You gave him the day off and now he is fired."

What the hell is he talking about? I thought. *What have I got to do with this?*

I stopped dead and forced myself to look at the ground. I was aware that people were watching. I had a warmth growing in me and I knew that I wanted to smash this man into oblivion.

He carried on yelling abuse and filth.

"I must not hit him," I said silently to myself but when I looked at him and saw how pathetic he was, saw the spittle running from the corner of his mouth and his arms flaying around in the air, all the violence left me, and I had one more try at saying something.

"Look, Mr Haughton…" I managed to get in before he burst into fresh vehemence.

"Look! Who the fuck are you to say look to me?" He screamed. Someone behind him shook his head and walked off.

"I am a senior official, I was mining before you were born. I am going to smack your stupid fucking face…"

I turned and walked away. I heard him shouting what a coward I was and how he would see me on Monday and so on, and so on.

As most of my people, with the exception of the three from Malawi, were processed I drank my coffee and went home to sleep.

I did not return on Sunday but Jacque called at the house in the late afternoon to say it was all over and that Hamilton-Walker had asked him to relay a message to me.

"We must report to work as normal tomorrow whilst the rest of the mine does nothing." He sighed.

"But what for?" I asked puzzled. "If there are no blacks, what are we going to do?"

"He has volunteered all of the whites, not production, but the fitters, planners and so on for one or other clean-up operation in the hostel."

"Oh shit," I replied. "I don't need any more of this bloody drama in my life. Do you know what that bastard Haughton did yesterday?"

Jacque nodded. "Brice told me. He said he was watching and if he had been you, he would have hit him."

"I couldn't do it, Jacque, there was too much going on and the man was pissed as a rat."

Jacque shook his head. "You should report him, he has been warned before. Anyway, it's not my business. He made an arsehole out of you, not me."

"Thanks," I replied.

"I would have hit him," he repeated.

On Monday, I turned up at work for one of the great lessons in my life.

I have, in my cynical fashion, developed a philosophy to cover trust in your fellow man. It clearly states: 'You must trust everyone'.

However, if this trust is betrayed, you must not be surprised. To trust takes faith, to accept betrayal calmly takes charity. On this awful Monday, I learnt much of the man in the street.

We arrived at the shaft and all grouped together in Mr Hamilton-Walker's office.

"Right," he started. "I am sure that I do not need to tell you about the events of the past few days, every one of you was involved at the stadium in one way or another."

Some of us nodded, others just waited.

He looked around and then continued, "On Friday evening, a lot of very good and trusted black employees of this mine were taken from their rooms and put into buses for the stadium. Now it was not our intention to punish innocent people but the workforce has to learn that when a union starts to flex its muscles, then the consequences can be painful. We don't want people not to support unions, we only want, as is reasonable, that they support sensible unions."

"So we now have a situation where there exists in the hostel thousands of rands of damaged and destroyed personal possessions. Possessions not only of the rioters but of the other seventy per cent, the little people who did not want trouble and tried to keep out of the way."

"Why are they damaged and destroyed?" One of the fitters asked.

Hamilton-Walker turned to him and said, "When we asked people to board the buses, the good ones did. The others who refused to listen, had to be found, caught and arrested. Whilst we were trying to do this, they managed to loot and destroy half the property left behind by the others."

He smiled. "When they got going, they really caused the sparks to fly. Anyway, now that the security forces have cleared the area, we can go in and clean up."

Harry interrupted. "Mr Hamilton-Walker, was the army present here on Friday? Was anybody killed?"

"The army was not here, in fact no one was here except the mine's own security forces. We handled the whole affair internally, although there was a strong police presence in Viljoensdorp waiting to intervene if the matter got out of our control. This is a big mine and we handle as many of our problems ourselves as we can. No one was killed to my knowledge."

Some small talk broke out and I heard mention of this friend of mine said this, old so and so told me…

Hamilton-Walker waited a few minutes listening to some of the comments. "Yes, I was in the hostel," he replied to a question.

Then he stopped us and said, "Harry, I want you to stay in the office and continue with the reports. This month will be worked on whatever portion is represented from the close of the last month until Friday. Oosthuizen, I want you to select two of the fitters to help you get our surface spares stored into some sort of order and be prepared to go underground to service the machines tomorrow. It is possible that this section will resume limited mining from next week if everybody is prepared to chip in and drive the machinery."

"Our section could possibly do very well as we are mechanised and not labour intensive. In fact, we could prove something here. The rest of you must report to the hostel manager who will divide you into teams and dish out tasks as it suits him. Before we bring in people to clear the mess, we want to salvage what few bits and pieces the bastards left behind."

When we were on the way to the hostel, I said to Jacque, "I thought you told me the production people were not going to be here."

He looked back at the small group of miners and said, "There are only a few of them."

Hendrik Brice was there with his miner 'Hakkies'. Also, a tall youngish man I did not know and Du Plooy, a coarse Afrikaner famous even among his friends as particularly stupid and a monumental persecutor of the blacks.

The hostel manager gave us a similar talk to what we had just had from Hamilton-Walker. He finished by dividing us into teams of three men each and assigning each team with a pair of bolt cutters, string, paper tags and packets of black plastic bags.

"When I have allocated you to different sections of the hostel," he said, "I want you to search systematically in each block in your area. Suitcases or sealed boxes must not be opened but they must be tagged with the room number and taken out to the grass area in front of each block. Then the lockers in the room that are closed or locked must be opened." He pointed at a bolt cutter. "That is what the bolt cutters are for. Everything of value in the locker must be put into a plastic bag, the plastic bag tied with string and a tag attached with the room number and the locker number on it."

He paused. "The idea is that when some of these people are re-employed, we will be able to identify and return their belongings."

Du Plooy snorted. "Half of the stuff we find will have been stolen in the first place."

I saw a flash of anger cross the manager's face and it was quickly stilled, he continued. "Some of these blacks have been with this mine for years and they deserve some consideration. I have especially asked that this exercise be undertaken and you people are the only whites available at such short notice to help me do it. Please try to save as much as you can and remember some of the stuff here could belong to your own boys. At lunchtime, I will provide you all with a good man-size steak from the kitchen." He stopped and smiled. "If I can find someone to cook it."

Most of us laughed and we moved off behind him, following as he sectioned our teams off to enter the hostel blocks.

I had not realised that the hostel was so immense. It was arranged in six small groups of buildings, each group had a central kitchen, which served two or three dining rooms. Around the outside of the building housing, the dining rooms and recreation facilities was a grass area cut by paths and small gardens. On the perimeter stood the triple storied blocks of rooms that housed the nearly thousand blacks allocated to each section.

I looked around as we walked. It must have been a pleasant place in which to live. That was no longer true, now it was a scene of desolation. There was not a single refuse bin standing upright; rubbish was strewn everywhere. We were passing a beer garden, in the middle of a sea of broken brown bottles was a pile of burnt furniture. The smell of flat beer and wet ash was strong and unpleasant. Hanging over balconies were torn curtains and items of clothing difficult to identify. All along the bottom of each block were mattresses and broken beds that had been thrown from the upper rooms to the ground below.

Doors had been torn from their hinges and everywhere you looked, broken panes of glass gave the buildings an abandoned air of finality. Gaping frames knew that nobody would ever live behind them again. Here and there amongst the discarded furniture were items obviously not issued by the mine. A small refrigerator distorted by its violent impact on the pavement lay amongst spoiling foodstuffs that had fallen from its shelves.

A sewing machine, and further along lay clothing, suits, jerseys and other items; splashes of colour on the ground. We walked past a knitting machine and some woollen baby clothes. A sign lay close by with a picture of small booties and a jersey underlined by the words, 'SEND SOME HOME FOR MAMA'S BABY!'. Someone's attempt to drum up business. How many enterprising hostel entrepreneurs had their small dreams shattered by this battle of the giants?

Then remembering the sports stadium, realised it had not been a battle of the giants but rather the slapping down of a snapping young cub at its meal by an infinitely stronger and more worldly-wise mother. It seemed to me that one day when we became old and tired, we in turn would be chased from the meat by something that was virile and hungry with little time for the weak and feeble.

Inside the rooms, the story was the same, lockers lay open, their contents strewn across the floors. Food and drink were on the walls and floors. Under the beds, we found an occasional box but nothing that still contained any item of value.

There was more or less a team for each group of buildings, approximately one hundred rooms per team. It looked like it would be a long job. Once the other teams had moved on and left the team I was with behind, it became lonely. There were only three of us: myself, DuPlooy and a fitter I knew as 'Red'. Red was also English and had lived most of his life in Rhodesia. I understood from the grapevine that he was an excellent diesel mechanic but was divorced and a little wild.

Duplooy told us to speak English as, although he was Afrikaans, he was quite happy in either language.

We stood on the bottom porch of the first block and looked around. The buildings stood silent and empty, the rubbish and broken furniture lay wasted.

"Shit," said Red. "It is fucking eerie, almost like the place thinks it is our fault and is about to leap out and swallow us."

DuPlooy looked around alarmed. "Do you think that there could still be some kaffirs hiding here? Maybe they didn't find them all, maybe we could get attacked."

Red laughed. "Yeah, you're right and if it was one of those last ones, you know, the 'Kaffirs' that want to kill all the Afrikaners, then you would be the first bastard he'd go for."

DuPlooy realised that he was being teased and squaring his shoulders, replied, "Well, he'd choose the wrong one because us Afrikaners know how to fix some dumb kaffir with a knobkerrie in his hand."

We started slowly because there was so much stuff all over the rooms and initially, we scratched around on top of the cupboards and in drawers, checking corners and shelves for whatever looked valuable. I started to realise that I was on the wrong wavelength when I saw Red pick up a penknife and examine it. He turned it over, opened and closed it and then put it in his pocket.

"Aren't we supposed to put valuables in the plastic bags?" I asked him.

"Of course, we are," he said. "If we know what fucking locker it comes from, but this…" He waved at the floor, "this could be anybody's stuff. It won't help if we start saving this."

I could not argue with this logic and did not quite know how to bring the penknife into it so I said nothing. In the room next door, DuPlooy started banging at something and then called out to us. "Hey, bring the bolt cutters, I've found a locker that's not been broken into."

Red and I went next door and we cut the padlocks on the door. I held the plastic bag open and the two of them emptied the locker into it. Some clothes, a few toiletries and some writing material.

Pete threw the last item into the bag and said, "Well, there is fuck all in there."

As I tagged and fastened the bag, I realised that these two intended to take anything of value for themselves. I was not prepared to let that happen.

The next closed locker we came to contained a leather photograph album, obviously expensive and holding about three full pages of pictures. They were typical family pictures; the only difference to the one I had at home was in the colour of the faces.

Red opened it and lifting the plastic on the first page started throwing the pictures to the floor.

"You can cut that out," I said. Both of them stopped and looked at me.

"I don't give a fuck what you think about the blacks or their property, whether they are animals or kaffirs or whatever, but we are not stealing what little chance they have of getting some of this stuff back."

Red shook his head. "Oh no, Royce," he said. "You don't stop me or anyone else. I have taken shit from the kaffirs all my life, all of us have. If we get a chance like this, I am taking what I can get."

I looked at DuPlooy, who shrugged. "He is right, man," he said. "Look, Royce, don't be stupid. These kaffirs don't believe that they will ever see any of this stuff ever again."

He turned to Red. "If it makes him feel better, throw the pictures into the bag."

Red threw two pictures into the plastic bag. "This album," he stated flatly, "is mine."

I dropped the bag and said, "Give it to me, Red."

DuPlooy took a step forward and stopped me. "Wait, wait a minute, this is bloody stupid. Here, Red." He stopped and picked up the bag. "Throw it in, man, it stinks of kaffir anyway."

Red hesitated and then threw the album into the bag and gave it to DuPlooy, saying, "You tag the bag and I will carry on to the next room."

I reached over and took the bag. "It's ok," I said. "I'll tag the bag."

DuPlooy shrugged and picked up the bolt cutters. Red glowered but followed him out of the room wordless.

When I had disposed of the bag, they were two rooms ahead. When I went in the room where they were, Red was there in a good mood.

"Ah, it's Sir Royce," he said and handed me a plastic bag. "Next door, Sir, we opened a locker in there, you can pack it away for your friends."

It was hopeless and I realised it. "I am going to report this, you guys are sick," I told him.

Red snorted. "Report what?" He asked. "The fucking album is in the bag, what else am I supposed to have coveted?"

"I'm reporting this anyway," I said and left them.

I walked out of the building and into the next, heading to the main administration block. I intended to see the hostel manager but hoped that I might run into Jacque. I wondered whether he was experiencing the same problems that I was. As I walked away, I looked up and saw one of the other miners, a man called Jan, above me on the first floor.

He shouted down, "Have you got bolt cutters?"

"No," I called up.

"Shit!" He looked up and down the length of the building. "Come up here and give me a hand," he said and disappeared.

I went up the stairs and walked along the balcony towards the rending and crashing noises that told me where he was. When I looked into the room he was on his own and struggling to force open a locker with an iron spike about half a metre long.

"Where are the others?" I asked.

"Oh, over in the next block," he replied, "but they are stupid, they call me stupid you know?" He looked around.

"Oh really," I said.

"Yeah, they do, but I am not." He picked up a small bag from the floor that clinked when he shook it. "See this? Whilst they are looking for bits and pieces, cooler bags and pens, all that shit, I'm going ahead and getting the bonus."

"What is the bonus?" I asked him.

"The money, man. See…" He opened the bag and I saw that the bottom was covered with silver and copper.

"The kaffirs left in such a fucking hurry that they forgot all the telephone change. I just force open the top of the door and feel inside the top shelf, that's where everybody puts their loose change. Here…" He gave me the spike. "You take whatever is in the next one."

"I don't want it," I said.

He was immediately surprised. "Why not? What is wrong? Are you too fucking rich?"

"No," I hesitated and then putting all the disgust I could muster into my voice and bearing, I continued, "I don't want something that some kaffirs been pawing."

I felt pathetic and somehow as dirty as they were, but it was lost on this man. He looked down at his bag and then at me, the statement made sense to him.

"Well," he said, "that's your bad luck." He shrugged. "All the more for me."

I left him with his bonus. Further on, I found Jacque carrying a cool box in one hand and a bolt cutter in the other.

"Hey, David," he called when he saw me. "Come over here."

I went over to him reluctantly and found what I knew I'd find. The cool box was looted, and its contents—a pen and pencil set, some leather gloves and various other small items—were looted.

"Aren't you ashamed of yourself?" I asked.

"Ashamed? Why should I be ashamed? What I don't take somebody else will take." I sat down on a chair standing amongst the debris.

"Oh, for fuck sake, David, grow up, man. Who does all this shit belong to anyway?"

"I don't know," I replied, "but it does not belong to you or anyone of us. It belongs to some nameless black that spent his life collecting it and then, because a bunch of shits like Skila Patio want to make a name for themselves has it…"

"First," I angrily put up one finger. "Jeopardised by a bunch of marauding so-called brothers…Second," I held up another finger. "Ignored by the management who is supposedly protecting him by whisking him off to the sports stadium, where the fucker is fired. Fired, you understand, and then finally…" I held up the last finger. "Has it stolen by someone who earns five times what he earns, someone who is supposed to be different and not do that sort of thing because, for fuck's sake, he is not a kaffir."

Jacque was, for once, at a loss for words. He stood and looked across the hostel parkland and said nothing. After a few minutes, he picked up his cool bag and said, "Maybe you should look over there," and pointed at a growing heap of black plastic bags. "You see even if some of us are picking up the odd item that is just lying around, we are still salvaging most of the main stuff. A lot of the blacks will get their stuff back."

I shook my head and got up. "I'm going to the office. I'm not having anything to do with this."

As I walked off, he called after me, "David, aren't you going to wait for the steak, the hostel guy said he would give us a steak for lunch."

On Wednesday, my three Malawis turned up. They had gone to Viljoensdorp on Friday afternoon and returned to realise from the activity that something was seriously wrong in the hostel. They had turned around and gone to a friend who lived in the native township. It did them no good. They had already been dismissed and could only collect their money and papers. They did not go back to Malawi but waited in the area.

Two months later, they heard that the nearest black recruiting office in a town about 50 kilometres from Consolidated Murchison was accepting applications from strays such as they were. As long as the mine had a clear record of behaviour against their company number and it could be checked, they would be re-employed. Most of the blacks who worked for our section returned to us over the next six months.

Michael is now a senior instructor at a large modern training centre for operators and Skila Patio returned but was arrested for intimidation in a later strike action. He is an expert at what he does and I am sure that he will be back to orchestrate a great deal more grief and frustration before he is finished.

I suppose that I was privileged to witness a little bit of history that has played its role in determining the course of the industry. These things must be seen in their greater context but for me, they can be weighed in the disappointment of an image reflected by a few discarded photos on the floor of a vandalised hostel room.

A Touch of Greatness

He was tall but slightly built, and although clean, sometimes wore uniforms that had been patched or had fraying cuffs and collars. He always arrived at his work earlier than he was required to do, but no matter how early he was, they were earlier. In the winter months, when breath comes from the body in great clouds of frayed vapour, they would be huddled and silent, disturbed only by a small ripple of satisfaction when the key grated in the lock of the outer door of his office. They had access to the surgery, a large draughty room, from the shaft side; whereas he had a private entrance from the car park outside.

He would enter and, watched by the rows of patient patients, take off his coat and hang it up, go to his desk and read the casualty list of the night before. Only then would he turn and, through the glass partition, assess the size of today's frenzy.

If you entered the surgery at any time during the next two or three hours, you would find him seated on a stool next to the counter. Behind the counter, one of the orderlies stood ready to search files, phone the hospital or help translate difficult symptoms. On his left was the head of a long line of men, each man ready to shuffle slowly forward as required until suddenly, he found himself before the intent and penetrating gaze of Skop.

The identity card carried by all mine employees is referred to as the 'Skop' card. William Coetzee, medical officer at 3 Shaft, Witkoppen Mine never discussed anything with any of his thousands of dependants unless they had this card in their hand.

The room was never silent, but the feel of the room spoke volumes; it was the combination of an assault on the senses, not the least of which included hacking coughs, sharp commands and inanimate rings, metallic clatter and trundling wheels. Behind it all was the word 'Skop'. Each man handled, had his interview terminated by this demand as the next man shuffled forward to take his place. William would lift his head to see who it was and then put out his hand. "Skop?"

Those in the line shuffled slowly forward, their morning regulated by the number of 'Skop' per hour.

So they called him 'Skop', this line of faceless men that queued every day to present a private grief. This line was never the same and yet always the same. The same problem with a different face, the same ignorance with the same result and yet never-ending, because it existed on a different penis, under a different foot, on a different arm.

Every day with the same care, Skop would open fetid dressings that had perhaps only remained in place because a shoe or a piece of stocking had not been taken off for a week or more. Every day, he would cut away filth and flesh, clean and dress, advise and admonish only to see the same on another shuffling body and then on another and another...

He knew that for them he was an evil necessity. He knew they distrusted him and would, if they could, rather walk the thousands of kilometres to their own specialists. Specialists who knew better than to destroy a man's pride by stripping his weakness before a room full of people. Specialists who understood the spirit of the body and talked to it through the casting of bones. Specialists who sucked the evil in tangible lumps from the flesh itself, curing the cause of the wrong and not just its symptom.

Skop knew nothing of these things...of the black man's medicine. He could only make the wound heal on the outside, leaving a fear inside; knowledge that somewhere the evil remained, potent and hungry.

But they, lacking the presence of magic in this white man's world of walls and rules, came and sat, coughed and listened. Each of them nursing the secret as long as they could until they could no longer and must bring the trouble to him. A man stands tall in the line and when his turn comes, does not speak but stares to the side as he hands over his card.

William calls the orderly. "Ask him," he says.

There is a rapid back and forth chatter of clicks and guttural utterances and the orderly says, "It is woman's sickness."

William motions at the man's trousers. "Open," he commands.

I look at this man and I see my presence disturbs him, but slowly he undoes a leather belt, unhooks some buttons and pushes his trousers down. His penis is black and swollen. It has an open sore the size of an egg on its side. Higher, amongst the pubic hair, there is evidence of yellow putrescence and it is horrifying. Willem takes a wooden spatula from a dish and lifts the penis, looking underneath. He nods and tells the man to dress. "Hospital," he tells the orderly and then says, "Enter his name and number in the book."

I know this book; it is the record of cases of sexually transmitted diseases. William has told me that he averages a hundred and twenty cases a month. All around the walls are posters advocating hygiene, some very explicit about sex.

At the door, the exit door, is a box full of condoms. They are free. I have seen some blacks take them but not many. The same ignorance, the same result, one man has learnt, or has he?

They shuffle forward in rhythm, secret, waiting.

They hate it and resent him.

He listens, exposes and prods, day in, day out. His pay is pitiful, his motivations suspect, but in this industry, his work essential. He does not believe himself to be great. Except, perhaps in a small way, he knows the role he plays and its scale, what he amounts to in their dark hidden world. That there is no greatness except that far off touch of knowing it.

Postscript

I have known James for a few years now; he is a strange and lonely man. He lives by himself on a boat surrounded by things acquired through impulse. Some artworks, a few books and a collection of music that ranges from jazz to country. He lost his wife to cancer and his children to an independence that he had instilled in them himself. He never had an understanding of permanence and therefore, never really felt he owned them in the first place. Everyone likes him but it is doubtful whether anyone loves him.

He is not a pack animal but does join the rest of us when it suits him. I think that he is, despite his cheerful ease with people, unwilling to open himself to friendship.

In recent years, he has taken to walking and has, on occasion, done this to an extent that has made him a fleeting public figure. He talks a lot and makes his living by working as a training consultant for anyone and everyone who will pay for the privilege of listening. He is considered something of a technical expert on mining by those who are supposed to know, so can from time to time find the work he needs to keep himself alive. Sometimes, just sometimes, you can find the passion in him.

That passion that joined him to the rock, blood and sweat of the environment that he loves. But even this is easily put aside for something else, something deep and callous, something I have never found in anyone else…

James has a sense of the moment and from that sense, denies the value of anything but the moment. If he ever loved at all, and he most definitely has loved, he loved his wife. She was his life above the ground. Her essence surrounded him and yet she too was almost lost to a moment of selfishness. Their relationship was saved even as it faltered through her death, which caught him up and swept him above the pettiness of his own concerns. James is always the gentleman.

But the moments of his life took from him the focus of his life; so he did not have what most of us built so carefully—a family, a sense of being important to others, an understanding of his role in society.

Once he told me, "I mourn my first wife often and I tell myself and my friends that it is because I loved her. But I think it is because she is a symbol. A symbol of much that I have thrown away, sacrificed to the passing importance of…well, you know, things…principles, schemes and ideas which went the same way she did."

What has not been thrown away or taken from him is the fitness of his life in the mine. I will never understand, and I doubt whether even he will understand, just what it was that allowed him to communicate so totally with men and machines, systems and rocks that surrounded him underground. But communicate he did and on a level that few artists and authors communicate with their work. James never went underground for pay as we all did; he went underground to find a life force that completed him, that made him whole.

He left the industry on a regular basis, any moment that came along became an excuse to deny that he was a miner, but he was always able to return. The mines were an entity that did not disappear because James had a moment. He told me the other day that he never went on leave…he just left.

Well, one day, he will just leave for the last time, as we all must do, and he will take with him his picture of the mines.

This book is a copy of that picture. It is a result of his understanding of the industry and the people that populate it. If you find in it a life that keeps bringing you back, you will have perhaps understood a little of the essence of the South African mining industry and James will have created a picture, something that will, finally, be more permanent than the moment.

James has remarried and has a young son. He is happy and in love. She accepts his history because it has made him who he is. He lives on a marina in Cambridgeshire but is now an old man and could not work underground if he wanted to.